T0245240

CAMBRIDGE LIBRARY COLLECTION

Books of enduring scholarly value

Life Sciences

Until the nineteenth century, the various subjects now known as the life sciences were regarded either as arcane studies which had little impact on ordinary daily life, or as a genteel hobby for the leisured classes. The increasing academic rigour and systematisation brought to the study of botany, zoology and other disciplines, and their adoption in university curricula, are reflected in the books reissued in this series.

Plants of New South Wales

Sir Ferdinand von Müller (1825–96) was a botanist renowned for his research on the native plants of Australia. After emigrating from Germany in 1847, he was appointed Government Botanist of Victoria in 1853 and subsequently Director of the Royal Botanic Garden, Melbourne, which post he held until 1873. He was elected a Fellow of the Royal Society in 1861 and was knighted in 1879 for his services to Australian botany. This volume, first published in 1885, contains Müller's botanical survey of the plants found in the Australian state of New South Wales. Including an introduction by prominent Australian botanist William Woolls (1814–93), the survey divides the flora into scientific orders, with short descriptions of genera and species. Both native and introduced plants are included in the survey. This volume offers valuable insights into the composition of Australian flora at the time of publication.

Cambridge University Press has long been a pioneer in the reissuing of out-of-print titles from its own backlist, producing digital reprints of books that are still sought after by scholars and students but could not be reprinted economically using traditional technology. The Cambridge Library Collection extends this activity to a wider range of books which are still of importance to researchers and professionals, either for the source material they contain, or as landmarks in the history of their academic discipline.

Drawing from the world-renowned collections in the Cambridge University Library, and guided by the advice of experts in each subject area, Cambridge University Press is using state-of-the-art scanning machines in its own Printing House to capture the content of each book selected for inclusion. The files are processed to give a consistently clear, crisp image, and the books finished to the high quality standard for which the Press is recognised around the world. The latest print-on-demand technology ensures that the books will remain available indefinitely, and that orders for single or multiple copies can quickly be supplied.

The Cambridge Library Collection will bring back to life books of enduring scholarly value (including out-of-copyright works originally issued by other publishers) across a wide range of disciplines in the humanities and social sciences and in science and technology.

Plants of New South Wales

*According to the Census of
Baron F. von Mueller ... With an Introductory
Essay and Occasional Notes*

FERDINAND VON MÜLLER
EDITED BY WILLIAM WOOLLS

CAMBRIDGE
UNIVERSITY PRESS

CAMBRIDGE UNIVERSITY PRESS

Cambridge, New York, Melbourne, Madrid, Cape Town, Singapore,
São Paolo, Delhi, Dubai, Tokyo, Mexico City

Published in the United States of America by Cambridge University Press, New York

www.cambridge.org
Information on this title: www.cambridge.org/9781108021050

© in this compilation Cambridge University Press 2010

This edition first published 1885
This digitally printed version 2010

ISBN 978-1-108-02105-0 Paperback

THE PLANTS

OF

NEW SOUTH WALES,

ACCORDING TO THE CENSUS OF

BARON F. VON MUELLER, K.C.M.G., F.R.S., &c., &c.,
GOVERNMENT BOTANIST OF VICTORIA.

WITH AN INTRODUCTORY ESSAY AND OCCASIONAL NOTES.

By WILLIAM WOOLLS, Ph.D., F.L.S.

SYDNEY: THOMAS RICHARDS, GOVERNMENT PRINTER.

1885.

3a 167—85

REMARKS ON THE FLORA OF NEW SOUTH WALES.

In considering the Flora of New South Wales as a portion of the Australian plants, now known to number nearly 9,000 species, it may be observed that many changes in the vegetation have occurred since the foundation of the Colony in 1788; that some plants have become rare in the localities in which they were first procured; that others once common in the neighbourhood of Sydney and Parramatta have disappeared before the progress of cultivation; and further, that species from various parts of the world, some introduced accidentally and some for industrial purposes, have taken the place of the primeval forests.

The Flora of New South Wales, therefore, has undergone great changes since the beginning of the century. It is not now what it was at the period when the illustrious ROBERT BROWN wrote his "*Prodromus Floræ Novæ-Hollandiæ et Insulæ Van Diemen,*" and it is certain that before another century greater changes may be anticipated.

In 1805 the whole population of New South Wales was little more than 7,000, and the immediate neighbourhood of Sydney was occupied by the primitive vegetation. Now, according to the last Census, the city and suburbs alone contain 250,000 inhabitants, and extend over 2,000 acres of ground, whilst substantial and in some instances splendid buildings, in 121 miles of streets, cover the area where the early botanists (especially Surgeon-General White) collected their specimens of Eucalypts and Proteads.

By the erection of buildings and the clearing of the land, not merely in Sydney and Parramatta, but even now beyond the Dividing Range (which was crossed by Wentworth, Blaxland, and Lawson in 1813, and opened a new territory to the colonists), there has been a great destruction of native plants, as well as a considerable introduction of foreign elements into the vegetation. It is estimated that between 700,000 and 800,000 acres of land are now under cultivation, and that cereals of various kinds, as well as many industrial plants, are rapidly displacing the flora of the past and establishing a different kind of vegetation; whilst some 170 species of exotics, principally obscure weeds or plants of little interest to the casual observer, are becoming naturalised in the Colony.

These circumstances are working changes of a very marked character in the settled districts, so that except in reserved forests, in gullies and mountainous places, or favoured spots set

A

apart for public purposes, the truly Australian plants are daily becoming less numerous. Nor is change limited to the eastern side of the Dividing Range, for the wonderful increase of sheep and cattle in the pastoral districts is exercising a marked influence on the Flora of the interior. It is calculated that the pastoral runs within the boundaries extend over 154 millions of acres, and that these runs are stocked with 2½ millions of horned cattle and between 35 and 36 millions of sheep. Now it is found that, as the sheep and cattle are perpetually feeding over the same runs (excepting in cases where they are removed from one paddock to another, or where in seasons of drought they are driven away to distant runs for pasture or water), certain plants on which they delight to feed are gradually disappearing from some parts of the Colony. This is especially true in regard to certain Salsolaceous plants, popularly termed "Salt-bushes," some of the native grasses once abundant, and even a few species of Acacia and Casuarina, which are eaten down as soon as they spring up. There is reason, therefore, to believe that in process of time sheep and cattle will occasion as marked a destruction of native plants in the pastoral as may now be noticed in the cultivated districts, and that graziers will find it necessary to utilise foreign grasses to supply the place of those passing away.

The wholesale destruction of Eucalypts, sometimes arising from natural causes, such as the ravages of opossums, insects, and fungi, or the unusual prevalence of storms and floods, but more frequently through the process of ringbarking, is another source of change. In the early days of the Colony, as soon as any one had obtained a piece of land, it was customary to set gangs of men to clear and burn off the timber on it without much discrimination. Trees of great economic value, as well as shrubby species of less utility, were thus wantonly destroyed, and consequently some of the farms near the towns first established were completely denuded of their trees, leaving scarcely any to shade the cattle from the scorching rays of the summer's sun, or to afford material for fencing, firewood, or rough buildings. Of late years ringbarking has taken the place of clearing and burning off by manual labour, and the Government encouraged private enterprise by allowing compensation for ringbarking at the rate of 1s. 3d. per acre. I am not now discussing the propriety or the impropriety of ringbarking generally as a means of promoting the growth of grasses; but it is a fact which cannot be overlooked that the destruction of large forests extending over many acres is altering the appearance of the country, affecting the marked features of its Flora, and probably influencing the climate for good or for evil.

Then, again, useful trees from all parts of the world, of which Baron F. von Mueller has treated so elaborately in his admirable work on "Select Plants," are being introduced for economic or ornamental purposes. These, though highly advantageous when planted in suitable localities for the production of timber, cannot fail in the course of time to efface the peculiarities of Australian vegetation. Whilst, however, the general character of the Flora may be changed, and in some areas plants now well known may perish, it is not to be supposed that in the uncultivated and wilder parts of the Colony native species will cease to exist. Great Britain, with an area of 121,000 square miles, which is rather more than a third of that of New South Wales, or 310,937 square miles, still retains in its less accessible districts, or in places reserved for various purposes in all their native features, some 1,600 species of flowering plants. Whilst, however, in Great Britain, certain plants have disappeared from the localities formerly assigned to them by Nature, and others now common are supposed to have formed no part of the primitive Flora, so also it must be with the plants of New South Wales; and therefore it may be reasonably expected that, as the clearing and cultivating of the soil extend, and as trees from different parts of the world take the places once occupied by Eucalypts and Proteads, the character of the vegetation will cease to resemble that of the geological period (miocene) with which it seems now connected, and will assume a very varied appearance.

As regards the Flora now existing, it may be seen that the plants of New South Wales have an intermediate character between those of Queensland and Victoria, and that whilst some of the genera in the Northern Districts are connected with India and China, there are in the Colony generally many of the true Australian type, and apparently endemic. For all Australia, Baron Mueller calculates that of the great orders, LEGUMINOSÆ, MYRTACEÆ, and PROTEACEÆ, the species are respectively 1,058, 651, and 586. Of the first of these orders, Western Australia reckons 38 genera and 439 species, and New South Wales 56 genera and 316 species. The genus Acacia, the most numerous of phenogamous genera in Australia, is represented by 122 species in Western Australia and 98 in New South Wales, only one species of which (A. *Farnesiana*) is common to the warmer regions of the world, and, though occurring in four of the Australian Colonies, it has not yet been found in Western Australia. It may be mentioned that the following genera are represented in New South Wales, and not in Western Australia :—Tephrosia, Wistaria, Sesbania, Carmichaelia, Glycyrrhiza, Zornia, Desmodium, Uraria, Lespedeza, Mucuna, Galactia, Vigna, Rhynchosia, Lonchocarpus, Derris, Sophora, Castanospermum, Cæsalpinia, Bauhinia, Barklya, Mezoneurum, and Neptunia.

Of these genera, New South Wales has 18, which are common to India and other parts; and hence, whilst this Colony possesses a large number of leguminosæ of the Australian type, it is also connected, especially in the warmer districts of the North, with the Flora of the East.

The great order of MYRTACEÆ affords another instance of the varied character of the Flora of New South Wales; for whilst of 651 species for all Australia, 380 are found in Western Australia and 140 in New South Wales, there are some genera common to India and this Colony and not represented in Western Australia. The genera not extending to the West are Tristania, Metrosideros, Backhousia, Rhodomyrtus, Myrtus, Rhodamnia, Eugenia, and Acicalyptus.

Six of these, especially those of the berry or drupe-bearing section, form another link with the vegetation of the East, for, as Mr. Bentham remarks, " The fleshy-fruited genera of the order are widely spread over the tropical regions both of the new and the old world, including many of the largest forest trees, and are in Australia almost limited to the tropics, a very few species extending into New South Wales, and only one into Victoria." The typical character of Australian vegetation is seen especially in the genus Eucalyptus; for though in the tertiary period it flourished in Europe, few species are now found beyond the continent of Australia. In Western Australia the species, especially those of the shrubby kind, are the most abundant; but it is remarkable that only one species of the larger kind, so far as yet known (E. *rostrata*), is common to all the Australian Colonies. E. *gracilis*, E. *uncinata*, E. *incrassata*, and E. *oleosa*, the species constituting the mallee-scrub, extend from the West to the arid parts of New South Wales, and give a peculiar character to the regions where they occur, but no species on this side of the Dividing Range is common to Western Australia. The Eucalypts of New South Wales form the most remarkable portion of the forest vegetation. None of them, however, excepting in the Southern ranges, approach in any degree the gigantic proportions of some species in Victoria and Western Australia, but they are widely distributed, and render the scenery truly Australian. Of the shrubby kinds, one (E. *obtusiflora*) is plentiful near the coast, and two (E. *stricta* and E. *stellulata* var. *microphylla*) are abundant on the elevated parts of the Blue Mountains. E. *Gunnii*, sometimes called "Swamp Gum-tree," which attains a height of 100 feet and more on the Mittagong Range, occurs in a dwarf state on the Snowy Mountains, and in company with E. *pauciflora* (E. *coriacea*, A. Cunn.), also much stunted in its growth, it has been found 5,500 feet above the sea-level. Eucalypts, therefore, in different forms adapt themselves to the arid regions of the interior and the snowy ranges of the mountains.

The PROTEACEÆ of New South Wales are represented by more genera but by a less number of species than in Western Australia; and it is singular that, whilst Dryandra is peculiar to the west, Stenocarpus, Lomatia, and a few species of other genera are found only on the eastern part of the continent. This order connects our Flora with that of Southern Africa; for although none of the species are identical, yet two of the tribes (Proteæ and Persoonieæ) are largely represented in Africa by Protea, Leucospermum, Mimetes, Sessuria, &c. But whilst the relation between the African and Australian Floras is of a tribal nature, one species of Persoonia extends to New Zealand, and Adenostephanus, Grevillea, and Stenocarpus are found in New Caledonia, the order being also represented in the Indian Archipelago and Japan, as well as in South America.

A consideration of the orders so largely represented in New South Wales may serve to give some idea of its peculiar Flora; but the Rev. J. E. Tenison-Woods well remarks, " New South Wales is not the best portion of the continent for studying the distinctive vegetation of Australia; in fact, there is a greater botanical difference between South-East and South-West Australia than there is between Australia and the rest of the globe." If, as Sir J. D. Hooker suggests, the typical character of Australian vegetation is fully developed in Western Australia, and that that region is to be regarded as a centrum from which plants have immigrated to other parts of the continent, it might naturally be expected that the Flora of New South Wales would be of a mixed kind, indicating that, in ages long past, the alpine portion of the vegetation had been connected with that of Tasmania and New Zealand. Hence, as Mr. Tenison-Woods continues, " The Flora is composed of an Australian character, which extends to Tasmania, and a semi-tropical one, which extends into Queensland." On looking at the map of this Colony it will be seen that, whilst great variety of climate is to be expected from its geographical position (seeing that it extends from 28° to 37° S. latitude, and 141° to 154° E. longitude, with a superficial area of 310,937 square miles), the natural features of New South Wales suggest that there must be some diversity in the vegetation. It is often remarked that there is something very monotonous in the forests of Australia, and that from the frequent occurrence of evergreen Eucalypts they assume a sombre and uniform appearance. This is, in a great measure, true; but it may be observed that the different regions of the coast, the mountain ranges, and the dry interior have many plants peculiar to themselves, and thus occasion a much greater variety than is generally supposed; whilst the geological formation of particular localities, especially where the trap has cropped up through the sedimentary rocks, is favourable for a luxuriant growth of plants. The eminent botanist, Allan Cunningham, in his trip over the Blue Mountains in 1823, and subsequently the far-famed geologists, the late Rev. W. B. Clarke and Mr. C. S. Wilkinson (the Government Geological Surveyor), noticed the remarkable change of vegetation resulting from this circumstance. The last observes—"From base to summit another range is clothed with a denser growth of vegetation than occurs elsewhere. The reason of this is that a rich chocolate soil has resulted from the surface decomposition of a basaltic trap dyke, which has burst through all the sedimentary rocks." This change of vegetation may be noticed particularly on Mounts King George, Hay, Tomah, and Wilson, and here and there in other parts of the Blue Mountains, where a similar formation prepares the soil for plants which do not occur on the sandstone or the Wianamatta shale. Cunningham noticed that on one spot on the way to Mount Tomah, Banksia *serrata*, Lomatia *salaifolia*, Isopogon *anemonifolius*, Lambertia *formosa*, and other Proteaceous plants were flourishing in their usual soil of decomposed sandstone, whilst a little further on, in the chocolate soil already mentioned, magnificent tree-ferns, gigantic climbers, epiphytic orchids, and trees different from those of the surrounding forests, were seen on every side.

In the coast region, as circumstances are more favourable for the growth of plants than in the interior, where long-continued droughts not unfrequently prevail, the species are comparatively more numerous and varied. It was from this region that R. Brown and the early botanists derived their specimens; and even now, in gullies, creeks, and uncultivated places, many of the beautiful plants which charmed the first collectors of them may be procured. Of the ten Eucalypts described in Willdenow's "Species Plantarum" (1799), nine were found in the immediate neighbourhood of Sydney; and though the hand of destruction has long been raised against them, they may yet be found in diminished forms between Sydney and Parramatta. The labours of Mr. R. D. Fitzgerald, F.L.S., in his elegant work on Australian Orchids (some of which were figured in 1813 by Ferdinand Bauer), clearly show that the numerous species of that order found by R. Brown in the vicinity of Port Jackson, in the early part of the century, still spring up in their appointed seasons, though some of them appear further inland than they were first seen. In the neighbourhood of Lane Cove, such plants as Doryphora *sassafras*, Quintinia *Sieberi*, Abrophyllum *ornans*, Hedycarya *angustifolia*, Alsophila *Leichhardtiana*, and Lomaria *Patersoni*, are yet to be found; and in more than one favoured spot on this side of the Dividing Range, Geijera *salicifolia*, Elæodendron *Australe*, Cargillia *Australis*, Aphanopetalum *resinosum*, Claoxylon *Australe*, Alchornea *ilicifolia*, Croton

Verrauxii, Cupania *semiglauca*, Nephelium *leiocarpum*, Euodia *micrococca*, and many interesting Cryptogams grow in all their native luxuriance.   There is reason to believe, however, that in all these cases, where the vegetation differs materially from that of the surrounding country, the underlying rock is some modification of trap.

On the banks of the Northern rivers and in the adjoining districts (particularly those of the Richmond and Clarence) some of our finest timbers and most interesting plants occur; and there Mr. Fitzgerald discovered several of the Orchids which had escaped the notice of previous observers.   To the south of Sydney, likewise, in the district of Illawarra, the favourable situation of the coast has given a semi-tropical character to the vegetation.   Ferns of various kinds, from the lofty Alsophila to the minute Trichomanes, luxuriate in abundance, and at least one species of palm (Ptychosperma *Cunninghami*) common to the northern part of Queensland, and also the " Cabbage Palm" of the colonists (Livistona *Australis*), which is gradually disappearing from the bays near Sydney.

The vegetation of the mountainous parts of New South Wales, though sharing many species in common with the coast districts, has its peculiar features.   On the Southern ranges some of the Eucalypts (especially E. *amygdalina* and E. *Stuartiana*) rise to the height of 150 to 200 feet; whilst the truly alpine plants common to Victoria and New South Wales are found equally in Tasmania.   Baron F. von Mueller enumerates forty-two such species, amongst which Stackhousia *pulvinaris* and Dichopetalum *ranunculaceum* are found 6,000 feet above the sea-level.   The same writer states that the following alpine plants are identical with European species :—Turritis *glabra*, Alchemilla *vulgaris*, Veronica *serpillifolia*, Sagina *procumbens*, Carex *pyrenaica*, C. *echinata*, C. *canescens*, C. *Buxbaumii*, Lycopodium *selago*, and Botrychium *lunaria*.

Veronica *densifolia*, a small, densely-tufted, much-branched, prostrate plant, occurs on the summits of Mount Kosciusko, and is nearly allied to the New Zealand V. *pulvinaris*.   Gaultheria *hispida*, also remarkable as being the only plant of the heath family in New South Wales, grows at an elevation of 4,000 to 7,000 feet on the Australian Alps, and at 4,000 feet on the Snowy Mountains, at the head of the Bellinger River; whilst the genus Aciphylla, which extends to New Zealand and still further south, is represented by A. *simplicifolia* and A. *glacialis* on the snowy summits of the Australian Alps.

As the Blue Mountains do not exceed 4,000 feet in height, the vegetation has not the same relation to the alpine plants of other countries as that of the Southern mountains.   At the Katoomba

cataract, however, Pherosphœra *Fitzgeraldi*, a diminutive plant of the Cypress kind, connects the flora with that of Tasmania. In the same category may also be mentioned Pennantia, Celmisia, Quintinia, Caladenia, and Todea, which have species common to both; whilst the beautiful tree-fern Dicksonia *Billardieri* is distributed through Queensland, Victoria, Tasmania, and New Zealand. Amongst the species as yet only known from the Blue Mountains the following may be enumerated :—

Epacris *reclinata*	Velleia *perfoliata*
E. *rigida*	Goodenia *decurrens*
E. *coriacea*	Acrophyllum *venosum*
E. *apiculata*	Pultenœa *glabra*
Monotoca *ledifolia*	Acacia *asparagoides*
Zieria *involucrata*	A. *gladiiformis*
Boronia *microphylla*	A. *obtusata*
Phebalium *lachnoides*	Persoonia *chamœpithys*
Asterolasia *buxifolia*	P. *revoluta*
Grevillea *laurifolia*	P. *mollis*
G. *Gaudichaudi*	P. *angulata*
G. *acanthifolia*	Caladenia *Nortoni*
Alania *Endlicheri*	Cryptostylis *leptochila*
Atkinsonia *ligustrina*	Lyperanthus *ellipticus*

These plants may serve to show that on the Blue Mountains there are many species of great interest to the botanist, not merely as indicating alliances with those of other countries, but as illustrating the effect of geological formation. A reference, however, to the maps of the late Revd. W. B. Clarke and Mr. Wilkinson confirms the sameness of the vegetation in general; for whilst the Wianamatta and Hawkesbury rocks cover extensive areas of the Eastern ranges, the basaltic rocks crop up only here and there, and occasion that rich harvest of species to which I have already referred.

In the depressed regions of the Western interior, though in favourable seasons numerous grasses, some other Glumaceæ, and flowering herbs spring up and cover the plains, the vegetation differs materially from that of the sea-coast or mountains. Amongst the trees, the Eucalypts (excepting those growing on the banks of the rivers) are for the most part stunted, and in some districts constitute in various species the "mallee scrub," whilst the most widely-distributed families are represented in the following manner :—SAPINDACEÆ, by Atalaya *hemiglauca*, Heterodendron *oleifolium*, Dodonœa *petiolaris*, D. *lobulata*, D. *viscosa*, and D. *boronifolia*; CASUARINEÆ, by Casuarina *stricta*, C. *glauca*, and C. *Cunninghami*; LEGUMINOSÆ, by Acacia *pendula*, A. *homalophylla*, and various species of Swainsona, Bossiœa, Templetonia, &c.; CAPPARIDEÆ, by Capparis *lasiantha*, C. *Mitchellii*, C. *nobilis*, and C. *loranthifolia*; MYOPORINÆ, by Myoporum *acuminatum*, M. *deserti*, and M. *platycarpum*; whilst of the beautiful genus Eremophila, the following species seem to cheer the loneliness of

the desert, viz. :—E. *Bowmanni*, E. *oppositifolia*, E. *Sturtii*, E. *Mitchellii* (the so-called "Sandalwood"), E. *longiflora*, E. *polyclada*, E. *bignonifolia*, E. *Brownii*, and E. *maculata*; SANTALACEÆ, by Santalum *lanceolatum*, Fusanus *acuminatus* (Quandong), Leptomeria *aphylla*, Exocarpus *spartea*, E. *aphylla*, &c. ; RUTACEÆ by Geijera *parviflora*; and PITTOSPOREÆ, by Pittosporum *phillyroides*. The pine tribe, or the CONIFERÆ, are confined to two species, Callitris *robusta* and C. *Endlicheri* ; but these are sometimes of considerable size, and available for industrial purposes. None of the RHAMNEÆ in the interior can be regarded as trees. The species which occur most frequently are Ventilago *viminalis*, Pomaderris *racemosa*, Spyridium *subochreatum*, Cryptandra *tomentosa*, and C. *amara*. Of the STERCULIACEÆ, of which so many species are limited to Western Australia, very few extend to the interior of New South Wales, the principal being Sterculia *diversifolia* (a tree rising sometimes to the height of 60 feet) and the shrubs Lasiopetalum *Baueri*, L. *Behrii*, Rulingia *pannosa*, &c. Codonocarpus *cotinifolius* of the PHYTOLACCACEÆ is a remarkable tree, occurring on pine plains or sandy scrubs ; and, according to Baron von Mueller, attaining sometimes a height of 40 feet. Sir Thomas Mitchell, in his " Expeditions" (vol. ii, p. 121), speaks of it as a rare tree, singular in appearance, and foliage tasting strongly of horse-radish.

In a region so extensive as that from the Dividing Range to the districts lying beyond the Darling there is considerable diversity of vegetation, arising from the difference of soil, the average rainfall during the year, and the comparative distance from any great river. There are some parts, for instance, in which, in good seasons, the native grasses and herbs suitable for pasture are very abundant. And then, again, there are parts where plants of the SALSOLACEÆ, AMARANTACEÆ, CRUCIFERÆ, UMBELLIFERÆ, and GERANIACEÆ (known by the popular names of "Salt-bushes," "Cresses," " Carrots," and "Crowfoots") seem wonderfully adapted to the nature of the soil, and afford nourishment for sheep and cattle when the ordinary grasses fail. The plains on which such herbage prevails consist for the most part of a red and chocolate loam ; but, as Baron Mueller observes in reference to the wide depressed interior, it is characterised " by subsaline or grassy flats, largely interspersed with tracts of scrub, and occasionally broken by comparatively timberless ranges. The great genus Acacia sends its shrubs and trees also in masses over this part of the country, where with their harsh and hard foliage they are well capable to resist the effect of high temperature during the season of aridity, while they are equally contented with the low degree of warmth, to which during nights of the cool season the dry atmosphere becomes reduced. . . . . Salt-bushes in great variety stretch far inland in this part of

Australia." In seasons of drought, when the annual rainfall does not reach 10 inches, vast tracts become destitute of herbage, and even in seasons comparatively good the vegetation, owing to the less quantity of rain which reaches the interior, forms a striking contrast with that of the coast districts. Our eminent Astronomer and Meteorologist, Mr. Russell, has afforded the means of estimating the effects of climatic changes on the Flora of New South Wales. In his invaluable work on the "Physical Geography and Climate of the Colony," he tells us that "the rainfall along the coast districts is very abundant, ranging from 45 inches at Eden to 70 at the Tweed River, in the extreme north. At Sydney it is 50 inches. Along the top of the mountains the rainfall is from 30 to 40 inches, on the Western slopes from 20 to 30 inches, and over the flat country from 10 to 20 inches." The temperature also, as recorded by the same writer, is very suggestive, for he states that, "whilst the mean temperature during the hottest months is 79·1 at Sydney and 82·6 on the Blue Mountains (Mount Victoria), it is 92·2 at Bourke and 94·1 at Wentworth, and that a considerable part of the Colony forming the Western plains is subject to greater heat, caused, no doubt, by the sun's great power on treeless plains and the almost total absence of cooling winds, the temperature there frequently rising over 100·, and sometimes up to 120· in the shade." The meteorological observations made by the able and distinguished astronomer, Mr. John Tebbutt, F.R.A.S., at his private observatory, Windsor, are also eminently calculated to show the wide range of temperature and great variability of rainfall on the Hawkesbury. These facts may serve to show the causes which for ages past have tended to characterise the Flora of New South Wales in the arid interior; but it is reasonable to suppose that, as the progress of pastoral pursuits yields gradually to the irrigation of the soil and the planting of forests wherever practicable, a new order of things will arise, and that the vegetation will be much altered in its essential features. Recent investigations of the fossil Flora of New South Wales and Victoria plainly indicate that the vegetation of this Colony was very different in ages long past from what it now is, and that species common to Europe in the Tertiary period are now found in a fossilised state in this Colony; whilst, in the opinion of Baron Mueller, to whom we are indebted for describing and figuring some relics of our former vegetation, the prevalence of certain species seems to indicate great climatic changes. The Flora of the past, therefore, is widely different from that of the present; and as time rolls on the Flora of the future will assuredly be associated with new elements. "*Tantum ævi longinqua valet mutare vetustas!*" Whilst then we are reminded that change is impressed on everything in this world, it is well to bear in mind that there is One who changeth not, and Who, in the infinity of His wisdom, reveals Himself from generation to generation in the adaptation of species to the altered circumstances in which they are placed.

In the enumeration of plants, arranged according to their respective orders, I have followed the system indicated in Baron Mueller's census, (1) because it was most convenient for me to do so in selecting the species from his work, and (2) because the Baron's extensive knowledge of Australian plants (which has been the result of thirty-eight years' residence in South Australia and Victoria, and of his various expeditions in other parts of the continent) entitles him to the greatest consideration in the classification of genera and species. There is a difference of opinion amongst eminent botanists as to the best method of arranging the orders, and therefore I feel that it would be presumptuous in me to decide whether the original system of De Candolle retained in the "Flora Australiensis" or the simplified and somewhat altered form of that system elaborated by the learned Baron is the better. Mr. F. M. Bailey, F.L.S., in his "Synopsis of the Queensland Flora" (pp. 890), and Mr. C. Moore, F.L.S., in his "Census of the Plants of New South Wales" (pp. 97), (both of which works are indebted in a greater or a less degree to the labours of Bentham and Mueller) have followed the arrangement of the Flora ; and it cannot be denied that many who have studied native plants prefer the system to which they have been accustomed. There is, however, much to be advanced in favour of the Baron's views, and men of candour and independent thought will not cling to a system simply because it is time-honored. The great difficulty is with the proper position of the Monochlamydeæ, or those plants which have only one floral envelope, as to whether they should stand by themselves in a separate series, or be incorporated, according to their respective alliances, with plants which are supposed to have two floral envelopes. In the system adopted by Mr. Bentham the Monochlamydeæ are placed separately in the fifth and part of the sixth volumes of the "Flora Australiensis," whilst, according to the views of the Baron, they should be distributed amongst the orders now arranged under the Thalamifloræ and Calycifloræ.

In the preface to his work on "The Native Plants of Victoria," Baron Mueller remarks that the system he has elaborated is that of the Candollean, or reversed Jussieuan, with the important change that the Monochlamydeous or Apetalous division is restricted to Conifers and orders closed allied to them. And then, after expressing an opinion that the system followed in the "Flora Australiensis" is partly artificial, not natural, so far as Monochlamydeæ are concerned, he adds—"To instance merely the disruption of affinities caused by holding separate the

Monochlamydeous series, and by adhering also too rigorously to
the different position of the fruit in relation to the calyx and
to staminal affixion, we have the extremely natural row of
amyliferous or curvembryonate orders distributed in the ' Flora
Australiensis' through all the five volumes of dicotyledonous
plants."

In order, therefore, to be consistent in this respect, such families
as the CARYOPHYLLEÆ, CHENOPODIACEÆ, FICOIDEÆ, &c., instead
of being separated as they are in the "Flora," should stand near
each other, seeing that they are nearly allied, not merely in the
shape of the embryo, but in other characters. The defects of
the system are further seen in the fact that, among the Thalamifloræ and Calycifloræ, which, strictly speaking, should have two
floral envelopes, no less than fifty-eight genera are entirely without petals, or contain species in which the corolla remains unde-
veloped merely so far as Australian plants are concerned, not to
speak similarly of extra-tropical vegetation. "Again," adds the
Baron, "in Haloragis, Myriophyllum, Cotula, Soliva, and some
other genera, the corolla is absent in the non-antheriferous
flowers, while even the very commencement of the Candollean
arrangement is made with the apetalous genera, Clematis, Thalic-
trum, and Anemone, soon to be followed by the equally well-
known Caltha. Furthermore, the EUPHORBIACEÆ, which in the
train of their alliances must carry with them always the URTICEÆ,
count amongst their numerous genera more than one-third provided with petals. Besides, in PROTEACEÆ, the floral envelope
may be regarded as homologous to that of the closely-allied
LORANTHACEÆ with an absence of a calyx, comparable to the
suppression of that organ in Diplolœna, Asterolasia, and few
other thus exceptional genera."

From these considerations, it is evident that the arrangement
hitherto accepted can be regarded only as provisional, and that
no system can be looked upon as perfect which separates orders
closely allied, or which, whilst professing to exclude from the
Thalamifloræ and Calyciforæ genera which have only one floral
envelope, places amongst them some which are monochlamydeous.
In the "Flora Australiensis," it must also be noticed that the
Gymnospermeæ are associated with the Monochlamydeæ, though
the former are destitute of any perianth; whilst in the two most
remarkable orders, Conifers and Cycads, the former show some
slight connection with club mosses and the latter have partly
the gyrate vernation of ferns. As regards the position of the
Casuarineæ, Mr. Bentham remarks that "the order is a very
distinct one; the floral structure may be nearly that of *Urticeæ*,
but the remarkable vegetative characters have no nearer parallel
than amongst some CONIFERÆ." Holding these views, the
CASUARINEÆ in the "Flora" stand next to the Urticeæ, and in

the same category with Coniferæ, whilst in the arrangement of the orders in the Baron's "Census" the CASUARINEÆ are placed, indeed, near the URTICEÆ; but as the species are not wholly destitute of floral envelope they are far removed from the Gymnospermeæ, in accordance with the views previously explained, being widely different also in the structure of the wood and in the full development of pistils with complete stigmatic apparatus.

The two great objects of Baron Mueller in arranging the natural orders have been to distribute the Monochlamydeæ in accordance with their respective alliances, and to place the Gymnospermeæ (the *Coniferæ* and *Cycadeæ*) by themselves, as he considers that no perfect natural system can be maintained without doing so. With regard to the restoration of some generic names and the suppression of others, which have become almost time-honored, I have deemed it advisable to give both of them (such, for instance, as Stylidium and Candollea, Ionidium and Hybanthus, &c., &c.); whilst, as the Baron is held responsible for the nomenclature generally, it has not been considered necessary in a work of a popular kind to give the individual authority for each indigenous species, as his aim was to adhere unexceptionally to clear priority, which he thinks is the only safe and lawful rule in scientific naming.

It remains now only to express my obligations to the Baron for the opportunity he has afforded me of extracting from his "Census" a correct list of the plants of New South Wales so far as they are yet known. Before the publication of the "Fragmenta Phytographiæ Australiæ," and the "Flora Australiensis," the preparation of such a list would have been impossible, and even with them the hand of the master was required to systematise the species scattered through the eleven volumes of the former, and to give a more correct view, so far as regards geographical distribution, of the plants described in the seven volumes of the latter. Baron Mueller has accomplished these objects for the Australian Colonies, and in so doing he has added to his world-wide reputation as the greatest of Australian botanists, whilst at the same time, though placed in a somewhat anomalous position of Government Botanist without a botanical garden, he has succeeded amidst many difficulties in tracing the development of native plants, in disseminating their seeds far and wide, and in elucidating their economic and medicinal properties. Those only who have studied these matters fully can appreciate the magnitude of the labours in which he has been engaged, or the benefits which his self-denying exertions have conferred on society.

PLANTS OF NEW SOUTH WALES.

I. DICOTYLEDONEÆ.

(I.) CHORIPETALEÆ HYPOGYNÆ.

The first family of dicotyledonous plants in the Census (the Choripetaleæ hypogynæ, or such plants as have disunited petals, or no petals, stamens inserted on the bottom of the calyx and at the base of the ovary, and the fruit free from the calyx) is that of the (1) RANUNCULACEÆ or Crowfoots, of which the common Buttercup may be regarded as typical. This order contains five genera, including seventeen species, of which the following occur in New South Wales, viz.:—Of the genus Clematis, C. *aristata*, C. *glycinoides*, C. *microphylla*, and C. *Fawcettii*; of Myosurus, M. *minimus*; of Ranunculus, R. *Millani*, R. *anemoneus*, R. *Gunnianus*, R. *lappaceus*, R. *Muelleri*, R. *rivularis*, R. *hirtus*, R. *parviflorus*, and R. *aquatilis*; and of Caltha, C. *introloba*. Clematis, or, as popularly termed, "The Virgin's Bower," is a genus of climbing plants, with pinnately or ternately divided leaves, and a profusion of white flowers. The most common form on this side of the Dividing Range is C. *aristata*, whilst C. *microphylla* is that of the interior. C. *Fawcettii* is a species on the Richmond River. Myosurus *minimus*, or "Mousetail," is a small plant only a few inches in height, and common not only to Victoria and Queensland, but also to Europe, Asia, and America. The flowers are very small and arranged in spikes on leafless stems. The species of Ranunculus, or "Buttercup," most frequent in the middle part of New South Wales, are R. *lappaceus*, R. *rivularis*, and R. *parviflorus*, whilst the rest occur for the most part in the Southern districts, as well as the introduced species, R. *muricatus* (Linn.). Caltha *introloba*, or "Marsh-marigold," as it is called in Europe, is a small plant with yellow flowers, radical leaves, and perennial habit, limited to moist places in the vicinity of the Australian Alps. Of the seventeen species of this order in Australia, therefore, fifteen are common to New South Wales.

2. The NYMPHÆACEÆ, or water-lilies, are floating plants not so common in the Southern as in the Northern hemisphere, and only two species, Cabomba *peltata* and Nymphæa *gigantea*, are known to extend to New South Wales; the former, with small purple flowers and peltate floating leaves, being found in ponds

and lagoons in the county of Cumberland; and the latter, conspicuous for its large showy flowers and reticulated leaves, in
lakes and marshes in the Northern parts of the Colony. Some
of the leaves measure 18 inches across, whilst the flowers have
sometimes above 200 stamens.

3. DILLENIACEÆ, or Dilleniads, are represented by eighteen
species of Hibbertia. These have yellow flowers, simple leaves,
and a shrubby habit, resembling in some respects the Buttercups
or Crowfoots, but from them differing in habit and properties.
The largest and most showy of the genus are H. *volubilis*,
occurring near the coast; H. *dentata*, on the banks of creeks;
and *H. saligna*, abundant on the mountains. Of the following
species, viz.: H. *nitida*, H. *bracteata*, H. *densiflora*, H. *stricta*,
H. *Billardieri*, H. *acicularis*, H. *Hermannifolia*, H. *vestita*, H.
serpyllifolia, H. *pedunculata*, H. *fasciculata*, H. *virgata*, H.
linearis, H. *diffusa*, H. *saligna*, H. *volubilis*, H. *dentata*, and H.
salicifolia, 14 occur in the county of Cumberland.

The section Pleurandra, formerly regarded as distinct, in con-
sequence of having the stamens all inserted on one side of the
pistil, is now incorporated with Hibbertia, as, according to
Bentham and Mueller, their characters appear much less important and less conformable to habit than was originally supposed. It is a remarkable fact that, besides the Australian
species, only two are known, and those from Madagascar, with
opposite leaves.

4. MAGNOLIACEÆ.—This order, which includes some of the
finest trees and shrubs in the world, is but poorly represented in
the Southern Hemisphere, and that by a genus somewhat
anomalous, viz., Drimys. D. *aromatica*, which is also common
to Tasmania, occurs in the southern extremity of this Colony,
being a mere dwarf in alpine situations, but rising in more
favourable places to the height of 30 feet. D. *dipetala* is a tall
shrub, found on the banks of creeks in many parts of the
Colony, and usually called "Pepper-tree" on account of the
pungency of its seeds. A third species has been discovered in
Lord Howe's Island (D. *Howeana*), and described by Baron
Mueller (*see* Frag., vol. vii, 17), but whether this is similar to one
known to occur near Rockingham's Bay is uncertain.

5. ANONACEÆ, or the order of the "Custard-apples," is for
the most part tropical, but of the species known to exist in
Australia, Ancana *stenopetala*, Polyalthia *nitidissima*, and Melo-
dorum *Leichhardtii* extend from Queensland to the northern
districts, whilst the genus Eupomatia (which deviates from
habitual structure in having perigynous stamens) has E. *Ben-
nettii*, from the sub-tropical parts of Australia, and E. *laurina*,

which has a wide range along the coasts of New South Wales. This remarkable plant was first described by R. Brown in the Appendix to "Flinders' Voyage."

6. MONIMIACEÆ (including the Atherospermaceæ of Lindley) have seven genera and fifteen species, limited for the most part to Queensland and New South Wales, two only extending to Victoria, and none to South or West Australia. Two trees, popularly known as "Sassafras" (Doryphora *sassafras* and Atherosperma *moschatum*), the one common near the eastern coast from Clarence River to Illawarra, and the other growing in the colder ranges of the south, are valuable for their industrial and medicinal properties. The genus Daphnandra is represented by D. *micrantha* of the Clarence, Richmond, and Lansdowne Rivers; Mollinedia by M. *Huegeliana*, from similar localities; Kibara by K. *macrophylla* and K. *pubescens;* Hedycarya, by H *angustifolia*, common to the Blue Mountains, as well as the northern and southern parts of New South Wales; Piptocalyx by P. *Moorei* from the Hastings; and Palmeria by P. *scandens* from Lane Cove and the Blue Mountains. When the 5th volume of the Flora was published there was some doubt about this species, as it was known only from imperfect specimens collected by the writer, but as Mr. R. D. Fitzgerald succeeded subsequently in finding flowering specimens of the same, Baron Mueller was able to identify it as P. *scandens*. This is not, as stated in the "Flora," a tree, but a straggling or climbing shrub with small flowers, and peculiar to Australia. It may be observed that the order Monimiaceæ is monochlamydeous, that is, the species have only one floral envelope.

7. The LAUREL FAMILY (LAURACEÆ) has here of Cryptocarya, C. *patentinervis*, C. *obovata*, C. *glaucescens*, C. *triplinervis*, C. *Meissneri*, and C. *Australis*. Of these, only C. *glaucescens* is found near Sydney, and it may be recognised by its black, shining, and aromatic berries. Nesodaphne or Beilschmiedia has N. *obtusifolia* from Clarence River; Endiandra, E. *Sieberi*, E. *discolor*, E. *virens*, E. *Muelleri*, and E. *pubens*, principally from the banks of the northern rivers; and Litsœa, L. *dealbata*, extending from the Richmond River to Illawarra. The leaves of this species are sometimes several inches long, smooth above and glaucous beneath, with rather prominent veins. Associated with these trees, but differing widely in habit, are the Dodder-like plants referred to the genus Cassytha. Of these, C. *glabella*, C. *pubescens*, C. *phœolasia*, C. *paniculata*, and C. *melantha* are found in different parts of the Colony. These plants are leafless, with filiform or wiry stems attaching themselves to living plants by means of suckers, and having an abundance of small drupaceous fruits, not unpleasant in flavour.

B

Like the true Dodders, they root in the earth, and as they become parasitical they lose their attachment to the soil, and sometimes passing from shrub to shrub become a tangled mass.

8. MENISPERMACEÆ.—This order has a climbing habit, inconspicuous flowers, and drupaceous fruit; and if a transverse section of the wood be made the medullary plates are seen to radiate from the central pith like the spokes of a wheel. The species of this Colony are Fawcettia *tinosporoides*, from the Richmond River; Carronia *multisepalea*, from the Bellinger and Clarence Rivers; and three species, Cocculus *Moorei*, Sarcopetalum *Harveyanum*, and Stephania *hernandifolia*, extending from Port Jackson to the Blue Mountains.

9. The POPPY FAMILY (PAPAVERACEÆ) is represented by a solitary species, Papaver *aculeatum*, which has flowers of a reddish colour, and is beset with prickles. The introduced plant Argemone *Mexicana* (Linn.) is becoming a nuisance in cultivated ground; and the common "Fumitory" (Fumaria *officinalis*), which is also referred to the poppy family, has established itself as a garden weed.

10. CAPPARIDEÆ.—The species of the Caper family do not occur near Sydney. Cleome *viscosa* is found on the banks of our northern rivers, Apophyllum *anomalum* on the Lachlan, and four species of Caper, Capparis *lasiantha*, C. *nobilis*, C. *Mitchellii*, and C. *loranthifolia*, in remote parts of the Colony. The first of these has an agreeable fruit, and the third, which is sometimes called the native orange, was first brought from the interior by the late Sir Thomas Mitchell. (*See* "Expeditions," vol. i, p. 314.)

11. The CRUCIFERÆ of New South Wales are not numerous near Sydney, and it may be doubted whether some of them are really indigenous. According to the views of Baron Mueller, the following species occur :—Nasturtium—N. *terrestre*. Arabis—A. *glabra*.   Alyssum—A. *minimum*.   Wilkia—W. *Africana*. Erysimum—E. *curvipes*, E. *brevipes*, E. *blennodioides*, E. *Cunninghami*, E. *Blennodia*, E. *capsellinum*. Geococcus— G. *pusillus*.   Menkea—M. *Australis*. Cakile—C. *maritima*. Barbarea—B. *vulgaris*.   Cardamine—C. *stylosa*, C. *dictyosperma*, C. *laciniata*, C. *hirsuta*, C. *eustylis*. Sisymbrium—S. *filifolium*, S. *trisectum*, S. *nasturtioides*, S. *eremigerum*, S. *Lucæ*, S. *cardaminoides*.   Stenopetalum—S. *velutinum*, S. *lineare*, S. *sphærocarpum*, S. *nutans*. Capsella—C. *elliptica*, C. *pilosula*, C. *cochlearia*, C. *humistrata*, C. *Andræana*. Lepidium—L. *leptopetalum*, L. *phlebopetalum*, L. *monoplocoides*, L. *papillosum*, L. *folicsum*, L. *ruderale*. Some of these, known as "Cresses," are edible; the "Shepherd's Purse" is common to Australia, Europe, and Asia; "Pepperwort" grows plentifully about cattle-stations; and "Sea-rocket" is found on the sandy shores of Eastern Australia, as well as on those of Europe. The introduced species

are Lepidium *sativum* (Willd.), Raphanus *raphanistrum* (Linn.), Sinapis *arvensis* (Linn.), Brassica *campestris* (Linn.), Sisymbrium *officinale* (Scop.), Senebiera *didyma* (Pers.), Capsella *Bursa-pastoris* (Moench), and Camelina *dentata* (Pers.)

12. VIOLACEÆ, or the violet family, has three species of Viola, similar to the cultivated violets, but deficient in scent— V. *betonicifolia*, V. *hederacea*, and V. *Caleyana*. The genus Hybanthus or Ionidium has a singular appearance, arising from the large size of the lowest petal. The flowers are blue or purple, and the leaves are either alternate or opposite. H. *Vernonii* and H. *filiformis* are frequent near Sydney, and H. *floribundus* and H. *enneaspermus* farther inland. Hymenanthera *Banksii*, though of the violet family, is a large shrub growing on the shady banks of the Nepean and other rivers.

13. The order FLACOURTIEÆ has in New South Wales Scolopia *Brownii*, Xylosma *ovata*, Streptothamnus *Moorei*, and S. *Beckleri*; but with the exception of the first, which extends to Illawarra, these species belong to the Northern Districts. The last two are glabrous twiners.

14. PITTOSPOREÆ.—This order is well known to gardeners as furnishing several species for shrubberies, one of which has scented flowers, and wood useful for xylography. Our species of Pittosporum are P. *rhombifolium*, P. *undulatum*, P. *bracteolatum*, P. *revolutum*, P. *phillyroides*, P. *bicolor*, and P. *erioloma*, but only the second and fourth occur near Sydney. P. *phillyroides*, from the interior, is a very graceful shrub with a drooping habit. Hymenosporum *flavum* is a handsome evergreen tree, and in the season its large yellow flowers are very conspicuous. Bursaria *spinosa* is a thorny shrub with a profusion of white flowers and purse-like capsules. Marianthus *procumbens* is a small heath-like shrub, with white flowers. Citriobatus *multiflorus* and C. *pauciflorus*, or the "orange thorns," are thorny shrubs, with fruit resembling diminutive oranges. Billardiera *longiflora*, B. *scandens*, and B. *cymosa* are shrubs with a climbing habit, greenish-yellow flowers, and edible berries; and Cheiranthera *linearis* is an under-shrub with showy blue flowers.

15. DROSERACEÆ, or the Sundews, are limited in New South Wales to the genus Drosera, comprising the following species :— D. *Indica*, D. *Arcturi*, D. *glanduligera*, D. *pygmæa*, D. *spathulata*, D. *binata*, D. *auriculata*, D. *peltata*, D. *Burmanni*, D. *Menziesii*. Many species are remarkable for the property which they possess of catching the small insects that alight on them, and in this respect they resemble D. *rotundifolia*, one of the insectivorous plants of which Darwin has written with so much interest.

16. The little order of ELATINEÆ, or Water-peppers, is limited to two small plants, Elatine *Americana* and Bergia *ammannioides*;

the former very common in watery places and marshes throughout the Colony, and the latter on the banks of the Darling and Murray, being also an Asiatic and African weed.

17. In the HYPERICINEÆ, Baron Mueller unites Hypericum *gramineum* with H. *Japonicum*. This little plant, popularly known as "St. John's wort," has orange-coloured flowers, numerous stamens usually united, and opposite leaves with pellucid dots. It is very widely distributed, and varies considerably in size.

18. POLYGALEÆ, or milkworts, have three genera in Australia —Salomonia, Polygala, and Comesperma. S. *oblongifolia* is rare in the Northern parts of the Colony, and is identical with the species found in India and the adjacent islands. P. *japonica* is also another plant which connects the flora of Australia with that of Asia. Comesperma is a genus truly Australian, with hairy seeds lengthened into a coma. Two of the species have blue flowers and a twining habit (C. *sphærocarpum* and C. *volubile*). The rest of the species are, for the most part, erect, shrubby plants—viz., C. *scoparium*, C. *retusum*, C. *ericinum*, C. *calymega*, C. *defoliatum*, C. *polygaloides*, and C. *sylvestre*.

19. The genus Tremandra, which gives its name to the order TREMANDREÆ, is peculiar to Western Australia; but of the genus Tetratheca one species, T. *ericifolia*, is common to all the Australian Colonies, whilst T. *juncea* has been found only in New South Wales. The species have purple flowers, four-celled anthers, and heath-like leaves.

20. The MELIACEÆ, of which the white and red cedars may be regarded as typical, comprise many fine trees, some remarkable for the utility and beauty of their woods, and others for their edible fruits. Trees known as cedars, rose-wood, yellow-wood, spotted trees, rose apple, sour plum, pencil-cedar, and pencil-wood belong to this order. Of these the red cedar is the most valued, rising sometimes to the height of 150 feet, with a proportionate diameter, and affording light, durable, and easily-worked timber. The species are more numerous in Queensland; but the following occur in New South Wales, especially in the Northern districts, the red and white cedars being amongst the few deciduous trees of Australia :—Turraea—T. *pubescens*. Melia — M. *Azederach*. Dysoxylum—D. *Fraserianum*, D. *Muelleri*, D. *Becklerianum*, D. *Lessertianum*, D. *rufum*, and D. *Patersoni*. Cedrela—C. *Australis*. Amoora—A. *nitidula*. Synoum—S. *glandulosum*. Owenia—O. *acidula*, O. *cepidora*. Flindersia—F. *Australis*, F. *Schottiana*, F. *Oxleyana*, F. *Bennettiana*, F. *Strzeleckiana*.

21. The RUTACEÆ contain some of our most esteemed flowers, and for many years past species of Boronia and Eriostemon have been prized in European conservatories. These plants, which

have for the most part hypogynous stamens, and fruit consisting
of several capsules, are generally characterised by resinous
pellucid dots and flowers of a white, pink, or blue colour. There
are 185 species recorded for Australia, of which about 80 occur
in New South Wales. Three of the genera are exclusively
Western Australian, and one tropical, whilst those which represent the orange or lemon family are limited to the Northern part
of the Colony or Queensland. The "Native Rose" (Boronia
serrulata) is known only from New South Wales, and some of
the most admired species of the genus belong to Western
Australia. Boronia *pinnata* is remarkable for having dimorphous
stamens, Eriostemon *myoporoides* for its very strong scent, and
E. *obovalis* for having occasionally double flowers. The genera
and species are thus re-arranged by Baron Mueller :—Zieria—Z.
*lævigata, Z. pilosa, Z. obcordata, Z. cytisoides, Z. Smithii, Z.
granulata.* Boronia—B. *algida, B. ledifolia, B. Fraseri, B. mollis,
B. microphylla, B. pinnata, B. cœrulescens, B. polygalifolia, B.
falcifolia, B. serrulata, B. parviflora, B. clavellifolia, B. Barkeriana,*
and B. *floribunda.* Eriostemon—E. *pungens, E. umbellatus, E.
Ralstoni, E. elatior, E. ambiens, E. rotundifolius, E. phylicoides,
E. ozothamnoides, E. Mortoni, E. sediflorus, E. lepidotus, E. alpinus,
E. squameus, E. ovatifolius, E. Beckleri, E. correifolius, E. Cun-
ninghami, E. mollis, E. trymaloides, E. Crowei, E. exalata, E.
lanceolatus, E. trachyphyllus, E. myoporoides, E. hispidulus, E. buxi-
folius, E. obovalis, E. scaber, E. linearis, E. difformis, E. ericifolius,
E. Coxii,* and E. *amplifolius.* Philotheca—P. *Australis.* Correa
—C. *alba, C. speciosa, C. Lawrenciana,* and C. *Baeuerlenii.*
Bosistoa—B. *sapindiformis,* B. *euodiformis.* Euodia—E. *penta-
cocca, E. Cunninghami, E. erythrococca, E. contermina, E. octandra,
E. micrococca, E. accedens, E. littoralis,* and E. *polybotrya.* Bou-
chardatia—B. *neurococca.* Xanthoxylum—X. *brachyacanthum* and
X. *Blackburnia.* Geijera—G. *salicifolia* and G. *parviflora.* Pleiococca—P. *Wilcoxiana.* Acronychia—A. *Baueri,* A. *lævis,* A. *melico-
poides,* A. *Endlicheri.* Halfordia—H. *drupifera.* Atalantia—A.
glauca. Citrus—C. *Planchonii,* C. *Australasica.* Pentaceras—P.
Australis. The genus Eriostemon, which was so called from the
woolly or hairy filaments of the species, is now extended by the
Baron to include several other genera which differ in some
measure from the original type. Some time-honored names
(such as Crowea and Phebalium) have thus been consigned to
oblivion; but the system of making large genera with groups
representing certain peculiarities is very convenient to the
memory. Many of the Rutaceæ are mere shrubs, remarkable
rather for the beauty of their flowers than for their industrial or
medicinal properties. A few, however, attain some size, and
afford useful timber. Geijera *salicifolia* and Acronychia *Baueri*
are of this character, the former having hard, closely-grained

wood, and the latter being somewhat similar. Pentaceras *Australis* is sometimes called scrub white cedar, whilst the wood of the lemon kind (Citrus and Atalantia) is closely grained and takes a fine polish. The medicinal properties of certain species remain to be investigated.

22. The order SIMARUBEÆ or Quassiads is for the most part tropical, and characterised by an intensely bitter principle. A few species extend to the Northern districts. These are trees or shrubs, with diœcious or polygamous flowers, drupaceous or capsular fruit, and bitter bark, viz.:—Ailanthus *imberbiflora*, Hyptiandra *Bidwilli*, Cadellia *pentastylis*, and C. *monostylis*. The genus of the last was made by Baron Mueller, and a figure of C. *pentastylis* is given in the "Fragmenta," vol. ii.

23. ZYGOPHYLLACEÆ, or bean-capers, are found principally in the Western interior, on the banks of the Lachlan, Murray, Darling, &c. They are, with one exception, herbaceous plants with opposite leaves, yellow or white flowers, and, in one species, with fruit valued by the aboriginal natives. (See Eyre's Expeditions, vol. ii, page 271.) Nitraria *Schoberi* is a rigid shrub, with small flowers, succulent leaves, and drupaceous fruit. In some countries the species of Zygophyllum are known to possess medicinal properties, but ours (Z. *apiculatum*, Z. *glaucescens*, Z. *iodocarpum*, Z. *ammophilum*, Z. *Billiardieri*, and Z. *fruticulosum*) have not been analysed. Tribulus *terrestris*, or Caltrops (so called from the form of its carpels), is a small prostrate plant, with pinnate leaves, yellow flowers, and prickly fruit. This is a common weed in the warmer parts of the world, and is troublesome to cattle by the prickly fruit running into their feet. The French call it La Croix de Chevalier. The form of the fruit resembles the machines which were cast in the way to obstruct an enemy's cavalry.

24. The FLAX family (Lineæ) is represented by the solitary species Linum *marginale*, a plant common to all the Australian Colonies, and differing from the northern L. *angustifolium* (D.C.) in the union of its styles. According to Baron Mueller, the bast of the Australian flax is of considerable strength, and well adapted for textile fabrics. The aborigines convert it into nets and fishing-lines. The seeds can be utilised for mucilaginous decoctions, and other purposes for which the common linseed is employed. This flax has blue flowers; but the introduced L. *gallicum* (Linn.), which is spreading very much on this side of the Dividing Range, has small yellow flowers.

25. Of the GERANIACEÆ, G. *carolinianum* or G. *dissectum*, G. *sessiliflorum*, Erodium *cygnorum*, Pelargonium *Australe*, P. *Rodneyanum*, and Oxalis *corniculata*, occur frequently in New South Wales. Geranium has usually ten fertile anthers and Erodium five, whilst Pelargonium is distinguished from both by its irregular flowers. E. *cygnorum* has blue flowers, and under the

name of "crowfoot"—a name more properly given to the buttercup—it is valued as a pasture plant. Oxalis *corniculata*, the
" sour-grass," is decidedly injurious when eaten in any quantity.
Erodium *moschatum* (Willd.), the European "musky heron's bill,"
has established itself in these Colonies, and is eaten by cattle.

26. The MALLOW family (MALVACEÆ), which may be distinguished from all other orders by its valvate calyx and monodel-
phous hypogynous stamens, is one largely distributed all over
the world, and comprises amongst its species many ornamental
flowers, as well as some of the commonest weeds, one of which,
Sida *rhombifolia*, has become troublesome in cultivated ground
in Queensland and New South Wales. The following species
are supposed to be indigenous:—Lavatera—L. *plebeia*. Mal-
vastrum—M. *spicatum*, M. *tricuspidatum*. Plagianthus—P. *pulchellus*, P. *spicatus*, P. *glomeratus*, P. *microphyllus*. Sida—S. *cor-
rugata*, S. *intricata*, S. *virgata*, S. *petrophila*, S. *subspicata*, S.
rhombifolia. Abutilon—A. *tubulosum*, A. *leucopetalum*, A. *micro-
petalum*, A. *cryptopetalum*, A. *otocarpum*, A. *avicennæ*, A. *oxy-
carpum*, A. *Fraseri*, A. *halophilum*, A. *Julianæ*, and A. *auritum*.
Pavonia—P. *hastata*. Howittia—H. *trilocularis*. Hibiscus—H.
rhodopetalus, H. *trionum*, H. *brachysiphonius*, H. *divaricatus*, H.
heterophyllus, H. *diversifolius*, H. *splendens*, H. *Kirchauffii*, H.
Farragei, H. *Sturtii*, H. *tiliaceus*, H. *tricuspis*, H. *insularis*.
Lagunaria—L. *Patersoni*. Gossypium—G. *Sturtii*.

It may be doubted whether Pavonia *hastata* and S. *rhombifolia*
are really indigenous, or like Modiola *Caroliniana* (Linn.), Malva
rotundifolia (Linn.), M. *parviflora* (Linn.), M. *verticillata* (Linn.),
and M. *sylvestris* (Linn.), of foreign origin; but amongst the
species truly Australia some are well worthy of cultivation,
especially the showy kinds of Hibiscus, Gossypium *Sturtii* (an
Australian cotton), and Lagunaria *Patersoni*, a pretty tree which
was introduced from Norfolk Island, but is now known to be
indigenous in Queensland and its southern borders.

27. Next to the Mallow family stands that of the STERCULIACEÆ,
in which the trees known as "Kurrajong," "Flame-tree," and
"Bottle-tree" are included. The flowers are generally small,
without petals, the stamens usually united into a ring or tube,
with five terminal teeth and the leaves frequently covered with
stellate hairs. According to the views of different botanists, the
species of Sterculia are variously arranged, some preferring to
unite Brachychiton and Delabechea with that genus, and others
retaining only two as true Sterculias. The latest division is as
follows:—Sterculia—S. *quadrifida*. Brachychiton—B. *discolor*,
B. *luridus*, B. *acerifolius*, B. *populneus*, and B. *populneo-aceri-
folius*. Tarrietia—T. *argyrodendron*, T. *trifoliata*. Ungeria—
U. *floribunda*. Melhania—M. *incana*. Commerçonia—C. *dasy-
phylla*, C. *rugosa*, C. *hermanniæfolia*, C. *Fraseri*, C. *echinata*.

Hannafordia—H. *Bissillii.* Seringea—S. *platyphylla,* S. *Hillii.*
Lasiopetalum—L. *dasyphyllum,* L. *Behrii,* L. *parviflorum,* L.
macrophyllum, L. *Baueri,* L. *rufum,* L. *ferrugineum.*

In this list Rulingia is incorporated with Commerçonia, and
Ungeria (as quoted Fragmenta, vol. ix, p. 169) is reckoned
amongst the Norfolk Island plants.

28. The TILIACEÆ, or Lindenblooms, are not numerous in
New South Wales. Grewia has one species (G. *latifolia*), a
plant nearly allied to the "plain currant," the seeds of which
Leichhardt found to impart an acidulated taste to water (see
"Journal," page 295); Corchorus has C. *Cunninghami,* one of
the plants remarkable for the toughness of its fibre; Sloanea, or
Echinocarpus (the "Maiden's Blush" of workmen), is represented by S. *Australis* and S. *Woollsii*; and Aristotelia by the
slender shrub A. *Australasica.* Elæocarpus (so called from its
olive-like drupes) is well known to cultivators by the species E.
cyaneus, which has loose racemes of pretty white flowers and dark-
blue drupes. E. *obovatus,* sometimes called native ash, is said to
have given the name to Ash Island. Of the other species, the
one is a tree of moderate size found on all the rivers of the north
(E. *grandis*), and the other a large tree (E. *holopetalus*) common
to the ranges of the south.

29. In the extensive order EUPHORBIACEÆ the species vary
from minute herbs to large trees, but they may generally be
recognised by their unisexual flowers, milky juice, and tricoccous
fruit. Though some of them are admired for their foliage, the
flowers are inconspicuous, and the wood not much esteemed.
Amongst the herbaceous species may be reckoned six of
Euphorbia, E. *Sparmanni,* E. *erythrantha,* E. *Drummondii,* E.
Macgillvrayi, E. *Norfolkiana,* E. *eremophila,* E. *obliqua*; two of
Monotaxis, M. *macrophylla* and M. *linifolia*; and three of
Poranthera, P. *ericifolia,* P. *corymbosa,* and P. *microphylla.* The
remaining species are either shrubs or trees:—Micrantheum—
M. *ericoides,* M. *hexandrum.* Pseudanthus—P. *pimeloides,* P.
ovalifolius, P. *divaricatissimus,* P. *orientalis.* Beyeria—B. *viscosa,*
B. *lasiocarpa,* B. *opaca.* Ricinocarpus—R. *pinifolius,* R. *Bow-
mani,* R. *speciosus.* Bertya—B. *gummifera,* B. *pinifolia,* B.
Cunninghami, B. *rosmarinifolia,* B. *Mitchelli,* B. *oleifolia,.* B.
*pomaderrioides,*B.*Findlayi.* Amperea—A.*spartioides.* Actephila
—A. *grandifolia,* A. *Mooreana.* Petalostigma—P. *quadriloculare.*
Phyllanthus—P. *Ferdinandi,* P. *thesioides,* P. *rigens,* P. *ramo-
sissimus,* P. *Gastræmii,* P. *subcrenulatus,* P. *microcladus,* P.
Fuernrohrii, P. *lacunarius,* P. *trachyspermus,* P. *Australis,* P.
thymoides, P. *Gunnii,* P.*filicaulis,* and P. *Tatei.* Breynia—B. *ob-
longifolia.* Hemicyclia—H. *Australasica.* Bridelia—B. *exaltata.*
Cleistanthus—C. *Cunninghami.* Croton—C. *insularis,* C. *pheba-*

lioides, C. *Verrauxii*. Claoxylon—C. *Australe*. Acalypha—A.
nemorum, A. *capillipes*. Adriana—A. *tomentosa*. Alchornea—
A. *ilicifolia*. Mallotus—M. *claoxvloides*, M. *Philippiensis*,
M. *discolor*. Macaranga—M. *Tanaria*. Baloghia—B. *lucida*.
Omalanthus—O. *populifolius*, O. *stillingifolius*. Excæcaria—E.
Agallocha. Tragia—T. *Novæ Hollandiæ*.

The trees of this order available for timber are Petalostigma,
Mallotus, Bridelia, Croton, Baloghia, " Scrub Blood-wood," and
Phyllanthus *Ferdinandi*, but they do not attain any great size,
and are chiefly used for cabinetmaking and carving, or for purposes in which durability is not essential. Some of the species
have barks containing medicinal properties, especially the so-
called Cascarilla (Croton *Verrauxii*), Petalostigma, and Bridelia ;
whilst others are remarkable for their acridity. Of the latter
class, Excæcaria *Agallocha*, or the " River Poisonous Tree," produces by incision in the bark an acrid, milky juice, which is so
volatile that nobody, however careful, can gather a quarter of a
pint without being affected. It is believed that a single drop
falling into the eyes will occasion the loss of sight. According
to Murrell's testimony, however, the natives of Cleveland Bay
used this poisonous juice to cure certain ulcerous chronic diseases.
This is a common maritime tree in the tropical parts of Asia,
and in Australia extends from the Gulf of Carpentaria as far
south as the Richmond River. Alchornea *ilicifolia* is the shrub
so celebrated for its parthenogenetic properties, having repro-
duced itself from seed in European gardens through several
generations from female plants alone. Many learned papers
have been written about this exceptional case, but the mode of
impregnation is still a mystery. This shrub is found on the
banks of our Northern rivers and also at Illawarra. Amongst
the species of the order introduced are Euphorbia *peplus*
(Willd.), or " Wartwort," common almost everywhere in gardens;
Ricinus *communis* (Willd.), found now on the banks of several
rivers in New South Wales ; and Euphorbia *lathyris* (Willd.), or
French Caper, in or near gardens, and known for its acrid and
narcotic qualities. It is said that the blacks use the juice of
a Euphorbia for sticking small feathers on native bees, in
order that they may be followed to their nests to obtain the
honey. This plant is E. *Drummondii*, a species very closely
resembling the European E. *chamæsyce* (Willd.), and not easily
distinguished from it. The Chinese are said to use the leaves
of Breynia *oblongifolia* or some allied species for adulterating
tea. On the whole, the order, though a large one, is rather
interesting for its industrial and medicinal properties than for
the beauty of its flowers or the value of its wood. It appears
that for all Australia 224 species are recorded, and of these more
than one-half are indigenous in Queensland.

30. The URTICACEÆ, or Nettleworts, which, like the preceding order, have only one floral envelope, are deficient in milky juice, and although in some instances they are large shrubs or trees, yet their wood is remarkable for lightness, sponginess, and profusion of cellular tissue. The leaves are generally covered with roughness or stinging hairs, and the fibre of some species can be utilised for cordage, sail-cloth, and textile fabrics. Two species of the nettle-tree (Celtis *paniculata* and C. *amblyphylla*) extend to New South Wales ; they are allied to C. *Australis* (Willd.), a European tree common in shrubberies, and known to children by its dark-coloured sweet drupes or fruit. Trema *cannabina* (under which name T. *aspera* and T. *orientalis* are united) is a tree of moderate size, and one of those which the aboriginal natives used to call "Kurrajong," as they employed the bark in tying. Aphananthe *Philippiensis* is a small tree, common to Australia and the Philippines. The flowers are monœcious and inconspicuous. Of the genus Ficus, or fig, the following species are found in different parts of the Colony, viz. : —F. *eugenioides*, F. *Muelleri*, F. *rubiginosa*, F. *columnaris*, F. *macrophylla*, F. *pumila*, F. *scabra*, F. *subglabra*, and F. *opposita*. F. *columnaris* is the Banyan of Lord Howe Island ; and the Moreton Bay fig, which is sometimes parasitical, has the singular property of enveloping and destroying large forest trees. Some years since Fraser remarked, " The roots of the Ficus increase rapidly, envelop the ironbark, and send out at the same time such gigantic branches that it is not unusual to see the original tree at a height of 70 or 80 feet peeping through the fig, as if itself were the parasite on the real intruder. In the singular angles or walls, as they are here termed, which are formed by the roots of these trees, and of which I have observed many 16 feet high, there is room enough to dine half-a-dozen persons." F. *pumila* or *stipulata* also at first has a prostrate or climbing habit ; but in course of time it throws out large branches, with leaves varying very much from the original ones. Cudrania *Javanensis* (the Morus *calcar-galli* of Cunningham) is a coarse straggling shrub, with strong thorns ; Malaisia *tortuosa*, called "Crow-ash," is of similar habit ; Pseudomorus *Brunoniana* is a kind of wild mulberry ; Elatostemma *reticulatum* and E. *stipitatum* are herbaceous or shrubby plants, without stinging hairs, and with long decurrent leaves ; Procris *montana* (Frag., vol. ix, p. 169) is a Norfolk Island plant ; whilst Boehmeria *calophleba* and B. *Australis*, the one from Lord Howe's Island and the other from Norfolk Island, are allied to B. *nivea*, from which the Chinese grass-cloth is made. Pipturus *propinquus* is a good-sized tree, with the underside of its leaves hoary. Parietaria *debilis* and Australina *pusilla* are small perennial plants ; Urtica *incisa* is a native sting-nettle, sometimes trailing to the length of 10 or 12 feet ; and Laportea *gigas* and L.

photiniphylla are the dreaded sting-nettle trees of Australia, rising to the height of 70 or 80 feet. The leaves of these are of a broadly cordate-ovate shape, sometimes more than a foot long, and so highly stimulating as to blister severely on the slightest touch. The common annual, Urtica *urens* (Linn.), is an introduced weed, occurring frequently in waste places and near buildings.

31. CUPULIFERÆ : This order in New South Wales is limited to a single species, Fagus *Moorei,* an Australian beech and an elegant tree, rising sometimes to the height of 150 feet, with a trunk of 70 feet to the branches. F. *Moorei* is hitherto known only from the mountain slopes near the Bellinger and Macleay rivers. Baron Mueller speaks of the allied species (F. *Cunninghami*) as being much used by carpenters and other artisans in Victoria and Tasmania.

32. CASUARINEÆ, or the Oaks of the colonists, have the following species :—C. *quadrivalvis,* C. *lepidophloia,* C. *glauca,* C. *equisetifolia,* C. *suberosa,* C. *Cunninghami,* C. *distyla,* C. *nana,* C. *inophloia,* and C. *torulosa.* Most of these trees and shrubs, which are known popularly as " Forest Oaks," " Swamp Oaks," " River Oaks," and " Dwarf Oaks," are diœcious, and the wood is utilised for shingles, axe-handles, and rough furniture. The branchlets of the shrubby kinds have a sub-acid flavour, and are relished by cattle.

33. The PIPERACEÆ, or Pepperworts, comprise Piper *excelsum* of Lord Howe's Island, a bushy shrub, sometimes attaining 20 feet; P. *Novæ Hollandiæ,* a tall dichotomous plant, climbing against trees in the dense forests of the North ; P. *hederaceum,* described by Cunningham as a magnificent woody climber, ascending to the tops of trees 150 to 180 feet high. The genus Peperomia is represented by the herbaceous plants P. *leptostachya,* P. *reflexa,* P. *Baueriana,* and P. *Urvilleana.*

34. Of the ARISTOLOCHIACEÆ, or Birthworts, only two—Aristolochia *prævenosa* and A. *pubera*—extend to New South Wales. These are climbing plants, with large perianths of a dull colour, and emitting an unpleasant odour. The order, according to Lindley, holds an intermediate position between exogens and endogens, agreeing with the former in essential points of structure, and with the latter in the ternary divisions of the flowers.

35. The order VINIFERÆ, or Vineworts, has, of the genus Vitis, seven species, viz.:—V. *Baudiana,* V. *nitens,* V. *acris,* V. *clematidea,* V. *hypoglauca,* V. *sterculifolia,* and V. *opaca.* Of these, V. *hypoglauca,* or the native grape, found in shady creeks and rocky places from the coast to the Blue Mountains, has edible berries, which make good jam and jelly. Attempts are being made in the South of Europe to cultivate this plant with a view of

counteracting the attacks of phylloxera vastatrix. V. *opaca*, as Leichhardt testifies ("Overland Expedition," p. 180), has edible berries and tubers, though of a pungent flavour. The latter contain a watery juice, most welcome to the parched traveller.

36. SAPINDACEÆ, or Soapworts, are composed of a great variety of species differing widely from each other, some being climbing shrubs and others forest trees. The flowers are usually polygamous and small. Atalaya *coriacea*, A. *multiflora*, and A. *hemiglauca* are small trees in the interior with pinnate leaves and panicles of white flowers; Diplottis *Cunninghami* is a much larger tree, found nearer the coast, with the habit and fruit of a Cupania and the flowers of a Paullinia; Cupania has C. *anacardioides*, C. *serrata*, C. *pseudorhus*, C. *xylocarpa*, C. *pyriformis*, C. *stipitata*, C. *tenax*, and C. *semiglauca*, for the most part small trees or shrubs, and with one or two exceptions limited to the banks of the Northern rivers; and Nephelium, N. *connatum*, N. *subdentatum*, N. *tomentosum*, N. *leiocarpum*, N. *Beckleri*, trees somewhat similar to Cupania. Heterodendron *oleifolium* and H. *diversifolium* are shrubs from the Western interior, with simple leaves and racemes of small flowers without petals. Harpulia, in its three species—H. *alata*, H. *Hillii*, and H. *pendula* (Tulipwood)—has trees of considerable size, with useful wood, whilst Akania *Hillii* is described as an elegant tree of 30 or 40 feet, with leaves 2 feet long, and long loose panicles of flowers. The genus Dodonœa consists principally of shrubs, with polygamous or diœcious flowers and hop-like fruits. With the exception of D. *triquetra*, which has hard closely-grained wood 3 to 4 inches in diameter, the other species are small, viz.:—D. *lanceolata*, D. *petiolaris*, D. *viscosa*, D. *peduncularis*, D. *procumbens*, D. *lobulata*, D. *truncatiales*, D. *triangularis*, D. *bursarifolia*, D. *Baueri*, D. *megazyga*, D. *pinnata*, D. *boronifolia*, D. *multijuga*, D. *adenophora*, D. *filifolia*, D. *tenuifolia*, and D. *stenozyga*. D. *cuneata*, common in the counties of Cumberland, Camden, &c., is united with D. *peduncularis*.

37. Rhus *rodanthema* and Euroschinus *falcatus* of the ANACARDIACEÆ are common to Queensland and New South Wales. They are small trees, exuding a caustic balsamic or gummy juice, the flowers inconspicuous, and the drupes fleshy. The wood of R. *rodanthema* is finely grained and beautifully marked, being much esteemed for cabinet work.

38. CELASTRINEÆ, or Spindle-trees, are represented by Celastrus *Australis*, a tall woody timber; C. *bilocularis*, a spreading tree; and C. *Cunninghami*, a small shrub; Leucocarpum *pittosporoides*, a tree with the trunk beautifully striated; Elæodendron *Australe*, a small tree conspicuous for its ovoid scarlet drupes; E. *curtipendulum*, lately added to our Flora; Siphonodon *Australe*, a tree of moderate size with a close-grained yellowish

wood; and Hippocratea *obtusifolia*, a tall woody timber, similar to a species common to the tropical parts of Asia. The flowers of this order are characterised by a large, expanded, flat disc, closely surrounding the ovary; Melicytus *ramiflorus* has recently been added to this order by F. v. M.

39. STACKHOUSIEÆ may be regarded as almost exclusively Australian, for only two species of Stackhousia are known out of this continent and Tasmania, viz., one in the Philippine Islands and the other in New Zealand. The genus consists of herbs with spikes of pentandrous flowers, and three minute bracts at their base. Baron Mueller reduces our species to S. *pulvinaris*, S. *linarifolia*, S. *muricata*, S. *viminea*, and S. *sphathulata*. Macgregoria *racemosa* extends to the North-western part of New South Wales.

40. The order FRANKENIACEÆ, which has the most species in Western Australia, has one common to all the Australian Colonies (Frankenia *lævis*), a shrubby plant, with sessile flowers and opposite leaves, found on the arid plains of the Darling, Murray, and Lachlan.

41. The small order of PLUMBAGINEÆ, or Leadworts, is found here in Plumbago *zeylanica*, a plant common to North Australia, Queensland, Tropical Africa and Asia, and the isles of the Pacific, and also in Statice *Australis*, common to Queensland, Victoria, Tasmania, New Caledonia, and Japan. The latter occurs along the coast in mud at high-water-mark.

42. The Purslanes (PORTULACACEÆ), which are most frequent in South America and the Cape of Good Hope, have in New South Wales Portulaca *oleracea* (a plant common in sandy places, and valuable for its succulent properties), P. *filifolia*, Claytonia *polyandra*, C. *pleiopetala*, C. *Pickeringi*, C. *volubilis*, C. *calyptrata*, C. *corrigioloides*, C. *pygmœa*, C. *brevipedata*, and C. *Australasica*. These are for the most part succulent herbs, growing in damp or marshy places. Only two are known to occur from Port Jackson to the Blue Mountains.

43. The CARYOPHYLLEÆ, or Chickweeds, are natives principally of the colder parts of the world, and therefore their numbers are comparatively few in New South Wales. The genera and species, including Scleranthus (the species of which are small densely-branched herbs most frequent in alpine regions), are—Stellaria *pungens*, S. *glauca*, S. *flaccida*, S. *multiflora*; Sagina *procumbens*, S. *apetala*; Colobanthus *Benthamianus*; Scleranthus *pungens*, S. *diander*, S. *biflorus*, S. *mniaroides*; Saponaria *tubulosa*; Spergularia *rubra*; Polycarpon *tetraphyllum*; and Polycarpœa *Indica*.

It may be doubted whether some of these are really indigenous. Stellaria *media* (D.C.), Cerastium *vulgatum* (Linn.), Silene *Gallica* (Linn.), and Spergula *arvensis* (Linn.), which are decidedly of foreign origin, are now very abundant in cultivated places.

44. AMARANTACEÆ, or the Amaranths, have three genera, widely dispersed over the warmer parts of the world, and four endemic. The species are principally weeds, though some of the genus Ptilotus, from the interior, are interesting plants. Alternanthera *triandra* and A. *nana* are glabrous weeds with opposite leaves; Achyranthes *aspera* and A. *arborescens* are herbaceous plants of little beauty and often spinescent; and Nyssanthes *erecta* and N. *diffusa* have small flowers in sessile heads, and troublesome thorny bracts. Trichinium being united with Ptilotus, the species are P. *hemisteirus*, P. *obovatus*, P. *parviflorus*, P. *alopecuroides*, P. *nobilis*, P. *macrocephalus*, P. *parvifolius*, P. *exaltatus*, P. *Manglesii*, P. *erubescens*, P. *Fraseri*, P. *sphathulatus*. Euxolus *Mitchellii*, E. *viridis*, and E. *macrocarpus* (differing little from Amaranths) are plants of little interest; Polycnemon *pentandrum* and P. *diandrum* are prostrate maritime herbs; whilst Deeringia *celosioides* and D. *altissima* are tall plants with spicate flowers and red berries. Amongst the introduced weeds of the order are several species of Amaranth.

45. The Salt-bush or Goose-foot family (SALSOLACEÆ) is a very important one in Australia, seeing that in dry seasons the species afford wholesome nourishment to sheep and cattle. In this family, such plants as "Salt-bush," "Cotton-bush," "Wild Spinach," "Roley Poley," &c., are included. Rhagodia (so called from its berry-like fruit) has R. *Billardieri* (a plant growing near the sea), R. *parabolica* (the larger salt-bush), R. *hastata* (the smaller salt-bush), and R. *Gaudichaudiana*, R. *crassifolia*, R. *spinescens*, R. *nutans*, and R. *linifolia*—all more or less available for pasture. The genus Chenopodium (including plants popularly termed "fat hen") has C. *nitrariaceum*, C. *auricomum*, C. *triangulare*, C. *microphyllum*, C. *carinatum*, C. *cristatum*, and C. *atriplicinum*. Dysphania *myriocephala* is a small annual remarkable for its numerous heads of minute flowers. Atriplex (the genus of the garden Spinach) has the following species widely scattered through the interior:—A. *stipitatum*, A. *nummularium*, A. *cinereum*, A. *rhagodioides*, A. *vesicarium*, A. *velutinellum*, A. *angulatum*, A. *semibaccatum*, A. *Muelleri*, A. *microcarpum*, A. *campanulatum*, A. *leptocarpum*, A. *limbatum*, A. *halimoides*, A. *holocarpum*, A. *spongiosum*, and A. *crystallinum*. The genus Kochia consists of small shrubs, amongst which are reckoned the "Cotton-bush," and its congeners Kochia *villosa*, K. *lobiflora*, K. *lanosa*, K. *triptera*, K. *oppositifolia*, K. *brevifolia*, K. *pyramidata*, K. *eriantha*, K. *sedifolia*, K. *humillima*, K. *microphylla*, K. *ciliata*, K. *brachyptera*, and K. *stelligera*. The genus Babbagia is endemic in Australia, and limited to a single species, B. *acroptera*; but under the name of Bassia (not derived from "Bass" of honored memory in Australia, but from F. Bassi, Curator of the Botanic Garden at Bologna), Baron Mueller has united nine genera with

the species B. *Dallachyana*, B. *scleroloenoides*, B. *diacantha*, B. *lanicuspis*, B. *bicornis*, B. *biflora*, B. *paradoxa*, B. *quinquecuspis*, B. *divaricata*, B. *bicuspis*, B. *echinopsila*, B. *enchyloenoides*, B. *brevicuspis*, and B. *salsuginosa*. Amongst these are "Roley Poley" and similar plants. Enchyloena *tomentosa*, Threlkeldia *diffusa*, and T. *proceriflora* are shrubs common to most of the Australian Colonies in the interior. Salicornia, or the " Marsh Samphire," has S. *robusta*, S. *arbuscula*, S. *leiostachya*, S. *tenuis*, and S. *Australis*, the last of which is very common in marshy places near the sea-shore. Suœda *maritima* is a plant of similar habit, with fleshy leaves, which can be utilised for pickling, &c. Salsola *kali*, found on sandy shores in many parts of the world, is burned for soda as used in the glass manufacture. The introduced species of the order are Chenopodium *murale* (Linn.) and C. *ambrosioides* (Linn.), common in gardens and waste places.

46. The FICOIDEÆ, or Pig-face order, has two species of Mesembryanthemum—M. *œquilaterale* and M. *Australe*, commonly called " pig-face" ; Tetragonia *expansa*, and T. *implexicoma*, improperly called New Zealand spinach; Aizoon *quadrifidum*, a rigid shrub with dichotomous branches; Sesuvium *portulacastrum*, a succulent herb, rooting at the joints, Zaleya *decandra* (Trianthema of the Flora Aust.) and Trianthema *cypseloides*, prostrate and diffuse herbs ; and Pomatheca *humillima*, described in the Frag., vol. 10, p. 72, as a minute plant from the Lachlan and Darling. The other species of the order are Macarthuria *Neo-Cambria*, Mollugo *Glinus*, M. *orygioides*, M. *spergula*, and M. *cerviana*.

47. The order of POLYGONACEÆ is well defined by the stipules cohering around the stem, and its usually triangular nut. Rumex or the Dock is represented by R. *Brownii*, R. *flexuosus*, R. *crystallinus*, and R. *bidens*. The species of Polygonum or Buckwheat are P. *plebeium*, P. *strigosum*, P. *prostratum*, P. *hydropiper*, P. *minus*, P. *subsessile*, P. *barbatum*, P. *lapithifolium*, P. *orientale*, and P. *attenuatum*. Most of these are common in moist or marshy places. Muehlenbeckia has M. *appressa*, M. *Australis*, M. *gracillima*, M. *rhyticarya*, M. *axillaris*, M. *polygonoides*, M. *stenophylla*, M. *Cunninghamii*. The first of these is a prostrate or climbing plant common in the Australian Colonies, and the last constitutes the " Lignum scrub," or Sturt's "leafless bramble" of the Narran, Darling, &c. Of this order some of the introduced species are becoming a nuisance in cultivated ground, such as Polygonum *aviculare* (Linn.), Rumex *crispus* (Linn.), R. (Mur.), R. *acetosella* (Linn.), and Emex *Australis*.

48. PHYTOLACCEÆ may be regarded as a small order nearly allied to chenopods and buck-wheats. The following species are indigenous, and occur beyond the Dividing Range. Monococcus *echinophorus*, Didymotheca *pleiococca*, Codonocarpus *Australis*,

and C. *cotinifolius.* The last of these, which occurs on the
deserts of the Lachlan, Murray, and Darling, is the most interesting, rising sometimes to the height of 40 feet, and distinguished
by its pale or glaucous green colour. An elaborate description
of this tree is given in Baron Mueller's "Plants Indigenous to
Victoria," p. 200. Phytolacca *octandra* (Linn.) has established
itself in many parts of the Colony, and is known by its dark-
coloured juicy berries.

49. Of the NYCTAGINEÆ, only three species extend to New
South Wales: Boerhaavia *diffusa* (of which there is a good
figure in Baron Mueller's "Lithograms"), Pisonia *aculeata,* and
P. *brunoniana.* The first of these has a long tap-root, resists
drought, and is a valuable pasture plant in the early spring, ere
the grasses are fully grown. The Pisonias are woody climbers
restricted to the northern parts of the Colony, but widely distributed over the tropical regions of the world.

From a review of the plants in the Baron's first division (that
is of such as have disunited petals, or no petals), it appears there
are forty-nine orders, 236 genera, and 710 species. It will be
seen that, differing from the system pursued in the Flora Australiensis, he proposes to incorporate the Apetaleæ of Jussieu,
or the Monochlamydeæ of De Candolle, with the petalliferous
divisions. Thus, amongst the plants enumerated, thirteen orders
or species with only one floral envelope have been associated with
the Thalamifloræ of vol. 1 in the Flora; whilst the FICOIDEÆ
of vol. 3, and the PLUMBAGINEÆ of vol. 4, from the alliance
of the former with PORTULACACEÆ, and from the fact of the
latter having petals sometimes free and the stamens hypogynous,
have been added to the same. In the preface to the Census the
Baron states his conviction that no perfect natural system can
be devised, so long as the Monochlamydeæ remain isolated and
associated with the Gymnospermeæ, and hence, in order to make
his arrangement as philosophical as possible, he places in his first
division of vasculares all plants which have divided petals or
perianths, hypogynous stamens, and fruit free from the calyx.

I. DICOTYLEDONEÆ.

(II.) CHORIPETALEÆ PERIGYNÆ.

The second section of dicotyledonous plants has for the most
part disunited petals; the petals, as well as the stamens, are
inserted on the tube of the calyx, and chiefly at a distance from
the base of the ovary, the fruit being laterally adnate to the
calyx, or, frequently in the LEGUMINOSÆ, free from it. This
arrangement comprises all the orders of the second volume of
the "Flora Australiensis" (with the exception of DROSERACEÆ),
the RHAMNEÆ of the first volume, and MYRTACEÆ, SALICARIEÆ,

ONAGRIEÆ, MELASTOMACEÆ, ARALIACEÆ, and UMBELLIFERÆ of the third volume. Thus, speaking in general terms, the first and second sections of the Baron's Dicotyledoneæ agree with the Thalamifloræ and Calycifloræ of Mr. Bentham, the monochlamydeous orders being distributed amongst them according to their respective alliances.

First of these orders comes that of the LEGUMINOSÆ, the largest order in Australia, and having for the whole continent upwards of 1,000 species. This order is divided into three suborders:—(1) Papilionaceæ (with pea-flowers); (2) Cæsalpinieæ (with regular or irregular flowers and free stamens); and (3) Mimoseæ (with small regular flowers sessile in spikes or heads).

PAPILIONACEÆ, FIRST TRIBE.

1. Stamens free (Decandra monogynia, Linn.) The genus Oxylobium consists of shrubs or under-shrubs with yellow flowers, the keel and base of the standard being of a darker colour. The following species are widely diffused throughout the Colony:—O. ellipticum, O. alpestre, O. cordifolium, O. Pulteneæ, O. hamulosum, O. scandens, O. procumbens, and O. trilobatum. The third of these is a minute species found only near the coast.

2. Chorizema is represented by the solitary species C. parviflorum, an under-shrub with orange flowers and linear leaves. The rest of the species are Western Australian.

3. Isotropis is also principally Western Australian, but one species, a broomlike shrub (I. Wheeleri), extends from South Australia to the western interior.

4. Mirbelia, a genus with yellow or purple flowers, has M. grandiflora, M. oxylobioides, M. reticulata, M. Aotoides, M. pungens, and M. speciosa. The species are small shrubs.

5. Of the genus Gompholobium, which is characterised by its club-shaped or broadly ovate pods and yellow flowers, some occur in the immediate neighbourhood of Sydney, and others from Port Jackson to the Blue Mountains. The species are G. latifolium, G. Huegelii, G. grandiflorum, G. virgatum, G. minus, G. uncinatum, G. glabratum, and G. pinnatum. The first of these is a shrub of some size, and in the spring its flowers are very showy. The last is a small plant, differing from the rest in its numerous leaflets.

6. Jacksonia is known principally in the neighbourhood of Sydney and the Blue Mountains by J. scoparia, a tall and almost leafless shrub (leaves being found only on young plants), broomlike appearance, and profusion of yellow or orange-coloured flowers. Baron Mueller has recently added J. rhadinoclada and J. Stackhousii to the species of New South Wales. No less than twenty-six species occur in Western Australia, besides J. scoparia.

C

7. Sphærolobium has one species common to four of the Australian Colonies and Tasmania. Most of the species belong to Western Australia. S. *vimineum* is a small plant, with slender leafless branches, terminal racemes of yellow flowers, and pods almost globular, scarcely two lines in diameter.

8. Viminaria *denudata*, common to all the Australian Colonies and Tasmania, is also an almost leafless shrub, sometimes rising to 20 feet, and abundant in marshy places near the coast. It has a broom-like appearance, with long racemes of yellow flowers, and small ovoid pods.

9. The genus Daviesia, though chiefly Western Australian, has in New South Wales D. *umbellulata*, D. *concinna*, D. *Wyattii*, D. *latifolia*, D. *corymbosa*, D. *filipes*, D. *squarrosa*, D. *ulicina*, D. *acicularis*, D. *brevifolia*, D. *pectinata*, D. *genistifolia*, D. *arborea*, and the small leafless D. *alata*, hitherto known only from Port Jackson. Some of the species are of considerable size, and on account of the bitter principle which pervades them they are called "hops." Horses and cattle are fond of browsing on the leaves. The flowers are yellow, orange, and red, and the pods triangular.

10. Aotus, a genus differing from Pultenœa in the absence of stipules and bracteoles, has three species common to New South Wales and Queensland—A. *villosa*, A. *mollis*, and A. *lanigera*. These are shrubs with simple hairy or villous leaves, yellowish clustered flowers, and flat ovate pods.

11. Phyllotus derives its name from the leaflike bracteoles inserted under the calyx; and the species are heathlike, with axillary or terminal flowers of a yellow or orange colour. P. *phylicoides* (under which several so-called species are included) is indigenous in New South Wales and Queensland. P. *humifusa* is peculiar to the southern parts of the Colony.

12. The species of Pultenœa are more numerous in New South Wales than in other parts of Australia. Though differing much in foliage and in inflorescence, yet they may be recognised by the persistent bracteoles under the calyx or adnate with its tube. The flowers are chiefly yellow or orange mixed with purple, and the stipules are linear-lanceolate or setaceous. The following, with one or two exceptions, are small shrubs :—P. *daphnoides*, P. *stricta*, P. *retusa*, P. *pycnocephala*, P. *polifolia*, P. *paleacea*, P. *Gunnii*, P. *scabra*, P. *Hartmanni*, P. *microphylla*, P. *pedunculata*, P. *ternata*, P. *stypheloides*, P. *altissima*, P. *obovata*, P. *incurvata*, P. *subumbellata*, P. *stipularis*, P. *glabra*, P. *dentata*, P. *aristata*, P. *plumosa*, P. *viscosa*, P. *echinula*, P. *hibbertioides*, P. *humilis*, P. *parviflora*, P. *procumbens*, P. *hispidula*, P. *villosa*, P. *foliolosa*, P. *flexilis*, P. *euchila*, P. *elliptica*, P. *subspicata*, P. *villifera*, P. *prostrata*, and P. *fasciculata* ; in all, thirty-eight species.

13. The genus Eutaxia, which differs only from the preceding in its decussate leaves and in the position of its bracteoles, has but one species (E. *empetrifolia*) in New South Wales, and that in the western interior. It is a glabrous and diffuse shrub, with yellowish flowers, and the keel deeply coloured.

14. Of Dillwynia, the species of which are somewhat variable, D. *ericifolia*, D. *floribunda*, and D. *juniperina* are common near Sydney, whilst D. *brunioides*, D. *cinerascens*, and D. *patula* are found farther inland. They are heathlike shrubs with yellow or orange-coloured flowers, scattered leaves, and small ovate pods.

PAPILIONACEÆ, SECOND TRIBE.

Stamens generally united in a sheath, or having the upper stamen free (Diadelphia, Decandria Linn).

15. Platylobium has only one species, P. *formosum* and its variety, *parviflorum*, in New South Wales. This is rather a handsome shrub with orange flowers, opposite heart-shaped leaves, and flat pods.

16. Bossiœa is a genus limited to Australia. The species are shrubs varying in size, in the position of the leaves, and in the colour of the flowers, some of which have a dark keel. B. *lenticularis*, B. *Kiamensis*, B. *foliosa*, B. *cinerea*, B. *prostrata*, B. *neo-anglica*, B. *buxifolia*, B. *rhombifolia*, B. *microphylla*, B. *heterophylla*, and B. *Scortechinii* are (with the exception of B. *prostrata*) erect or diffuse shrubs, with opposite or alternate leaves, yellow, orange, or red flowers, and flat pods. Some of these species flower in winter. B. *riparia*, B. *ensata*, B. *scolopendria*, and B. *Walkeri* are leafless shrubs, with the branches usually much flattened or winged.

17. Templetonia has in New South Wales T. *Muelleri*, with a few flat coriaceous leaves, and T. *egena* and T. *sulcata*, which are leafless. These are small shrubs, flowers varying from yellow to red, and very flat pods. Two of these occur in the deserts of the interior.

18. The species of Hovea found near Port Jackson are H. *linearis*, H. *heterophylla*, and H. *longifolia* ; H. *acutifolia* and H. *longipes* occur in the northern parts of the Colony. H. *longifolia* is a tall shrub, varying very much in the shape and character of the leaves. The flowers are blue or purple, usually in short racemes.

19. Goodia is a small Australian genus having one species in New South Wales (G. *lotifolia*). This is a tall shrub, with pinnately tri-foliate leaves, and flowers yellow streaked with purple. It occurs in shady ranges and on the banks of rivers, but is more abundant in Tasmania.

20. Crotalaria has the following species in the interior or northern parts of New South Wales:—C. *linifolia*, C. *retusa*,

C. *Mitchelli*, C. *incana*, C. *medicaginea*, and C. *dissitiflora*. The flowers are yellow, in terminal racemes, and the pods inflated. One of the most interesting species is C. *Cunninghami*. The flowers are very large, of a greenish colour, and the leaves are densely villous. This plant belongs to the hotter parts of Australia, and is difficult to cultivate.

21. Trigonella *suavissima* is an interesting plant, an annual, and of a prostrate habit, found on the Namoi, Darling, &c. Sir T. Mitchell (vol. i, 254) was the first to find this fragrant trefoil in the dry bed of the Darling, and commends it for its agreeable perfume and its delicious flavour. He regarded it as a new form of Australian vegetation, resembling that of the South of Europe. Mr. Bentham says that the species is very nearly allied to an Egyptian one. The flowers are yellow, in sessile clusters, and the leaves trifoliate.

22. Lotus may generally be distinguished by its five leaflets—three almost digitate at the end of the petiole or stalk, and two close to the stem. L. *corniculatus* and L. *Australis* (the one with yellow and the other with pinkish white flowers) are prostrate herbs, frequent in many parts of the Colony.

23. Psoralea (a genus which derives its name from the numerous little tubercles or glands on most of the species) is not represented by any species near Sydney. With the exception of P. *adscendens*, which occurs on the Mittagong range, most of the species, of which P. *eriantha*, P. *patens*, P. *cinerea*, P. *parva*, and P. *tenax* may be mentioned, belong to the river banks of the interior. The flowers are generally purple, the leaves digitate, and in some species the stems so tough as to be broken only with difficulty (Mitchell, vol. ii, p. 10).

24. Indigofera *australis*, a very variable plant, has a wide range in Australia. It has pink or purple flowers, pinnate leaves, and terete straight pods. The other species, I. *linifolia*, I. *enneaphylla*, I. *trita*, I. *hirsuta*, I. *brevidens*, I. *coronillifolia*, and I. *efoliata*, are, for the most part, northern or western plants, some of which are common to India and tropical America.

25. Tephrosia *purpurea* and T. *Bidwillii* extend to New South Wales, but the species are numerous (24) in North Australia. These are shrubs with racemes of purple flowers, pinnate leaves, and linear falcate pods.

26. Wistaria *megasperma* is a tall woody climber found chiefly on the Richmond River. It has racemes of purple flowers, pinnate leaves, and woody pods with large thick seeds. W. *australis* is also a native of New South Wales. Mr. Bentham refers these plants to Milletia, a large genus ranging over the warmer regions of Asia and Africa.

27. Sesbania *aculeata* resembles one of the tropical species of Asia and Africa. It has small yellow flowers, pinnate leaves, containing from twenty to fifty pairs of leaflets, and long narrow pods. It is found on the Darling.

28. Carmichaelia *exsul* is described by Baron Mueller as a small shrub from Lord Howe's Island (Frag., vol. vii, p. 126). Don gives C. *Australis* as common to New Holland and New Zealand, but the Baron regards C. *exsul* as distinct from any New Zealand species.

29. Clianthus *Dampieri*, or, as sometimes called, "Sturt's Desert Pea," is well known as a cultivated plant, and is remarkable as being one of those collected by Dampier near Shark's Bay in 1699. Some of Dampier's specimens are still existing in the Museum of Oxford, and from a figure of the plant in question Baron Mueller has identified the specimen as the beautiful C. *Dampieri* of Cunningham. (See Record of Plants collected by Mr. P. Walcott and Mr. M. Brown, 1861, &c., by F. von Mueller.)

30. Streblorrhiza *speciosa*, from Norfolk Island, described as a climber under the name of C. *carneus*, is now regarded as belonging to a distinct genus.

31. The pretty genus Swainsona, the species of which have been so widely cultivated, has the following in New South Wales, occurring for the most part beyond the dividing range :— S. *Greyana*, S. *galegifolia*, S. *coronillifolia*, S. *brachycarpa*, S. *phacoides*, S. *Burkittii*, S. *oligophylla*, S. *campylantha*, S. *plagiotropis*, S. *procumbens*, S. *phacifolia*, S. *oroboides*, S. *lessertiifolia*, S. *monticola*, S. *microphylla*, S. *laxa*, S. *Fraseri*, and S. *oncinotropis*. Some of these plants are poisonous to sheep and cattle, especially in dry seasons, when the grass fails. One is called the "poison pea" of the Darling, and has an extraordinary effect on the animals browsing on it by causing them to climb trees and to see objects larger than they really are.

32. Glycyrrhiza *psoraleoides*, which is well figured in Baron Mueller's Lithograms of Victorian Plants, is the only species of the genus known in Australia, and is limited to the banks of the rivers to the west of the Dividing Range. This is allied to the European species, and also to *Liquoritia officinalis*, the liquorice of commerce.

33. Zornia *diphylla* is a small decumbent herb, having orange-coloured flowers, two leaflets at the end of the petiole, and pods with several articulations. This is a plant common both to the new and old world.

34. Desmodium is a genus, so called from its jointed pods. The species are mere herbs. D. *brachypodum*, D. *varians*, and D. *rhytidophyllum* are common near Sydney, and, as the pods are covered with short clinging hairs, they are troublesome in adhering

to the dress. D. *acanthocladum*, D. *nemorosum*, D. *polycarpum*, D. *trichocaulon*, and D. *parvifolium* also occur in other parts of the Colony.

35. Uraria *picta* is an under-shrub common to India and Queensland, and also extending to the Northern districts of New South Wales. It has purple flowers, calyx with setaceous lobes, pinnate leaves, and articulated pods.

36. Lespedeza *cuneata* (which is found in North America, East Indies, and Archipelago) is a small plant with pinkish flowers, trifoliate leaves, and small pods nearly round. It spreads in cultivated ground.

37. The genus Glycine consists of small twining or prostrate plants, two of which—G. *clandestina* and G. *tabacina*—are very common. The other species, G. *falcata*, G. *sericea*, and G. *tomentosa*, belong to the interior.

38. Kennedya (including Hardenbergia) has four species in this Colony, K. *procurrens*, K. *rubicunda*, K. *prostrata*, and K. *monophylla*. The first three of these have trifoliate leaves and reddish flowers, and the last simple leaves and blue flowers. K. *procurrens* is limited to the warmer parts of New South Wales, but the others are found from the coast to the Blue Mountains. The species most in favour with cultivators are Western Australian.

39. Erythrina *Indica*, a prickly plant with large scarlet flowers, is common to the East Indies, North Australia, Queensland, and New South Wales.

40. Mucuna *gigantea* (a plant allied to M. *pruriens*, "the cowage" of chemists) is a coarse twiner, with greenish flowers and thick pods containing large round seeds.

41. Galactia *tenuiflora* is another twining plant, but much smaller than the preceding, with pale reddish flowers, leaflets three, or rarely one or five, and linear, flat, coriaceous pods. It occurs at Clarence River, and also in the interior.

42. Canavallia *obtusifolia* is found near the coast, and has prostrate or trailing stems. The flowers are pinkish, leaflets three, and pods (which are edible) somewhat broad, with longitudinal wings.

43. The genus Vigna is represented by the four species V. *vexillata*, V. *lutea*, V. *luteola*, and V. *lanceolata*. These are prostrate or twining plants, with trifoliate leaves, yellow or purple flowers, and linear pods. The first occurs on the Blue Mountains, and the others in the Northern districts.

44. Rhynchosia *minima* is a slender trailing plant, with trifoliate leaves and pendulous racemes of yellow flowers, apparently rare in New South Wales.

45. Lonchocarpus *Blackii* is a tall woody climber, with dark purple flowers, pinnate leaves, and pod very thin with broad, flat, reniform seeds, and, as the preceding, a Northern species.

46. Derris *scandens* resembles the last in habit, but with more numerous leaflets and racemes of yellowish flowers.

47. Æschynomene *Indica* extends from N. Australia to the Northern districts. It is a diffuse annual, with very numerous leaflets.

SOPHOREÆ, THIRD TRIBE: Stamens all free, or scarcely united at the base.

48. Sophora is a genus consisting of shrubs with numerous pinnate leaflets, yellowish or violet flowers, and chain-like coriaceous pods. The genus is widely distributed over the warmer regions of the old and new worlds. S. *tomentosa* and S. *Fraseri* are known from the Hastings and Clarence rivers, and S. *tetraptera* from Lord Howe's Island (Fragmenta, vol. vii, p. 26).

49. Castanospermum *australe*, or the "Moreton Bay Chestnut," is one of the finest of Australian trees, extending from the north of Queensland to Clarence River. The seeds of it, after due preparation, are eaten by the aboriginal natives.

50. Barklya *syringifolia* is a handsome tree of some size, occurring sparingly in the northern parts of the Colony. The leaves are simple, the flowers small and yellow, and the pod about 2 inches long, with few seeds.

SUB-ORDER 2. CÆSALPINIEÆ.

Flowers usually with five petals; stamens, ten or fewer all free.

51. Cæsalpinia *Bonducella* (Guilandina of Flora Aust., vol. ii, p. 276) is a shrub with spreading or climbing branches, armed with hooked prickles, flowers yellow in racemes, and leaves twice pinnate. It is allied to C. *sepiaria*, an East Indian species, much used for hedges, and ranges from Queensland to the northern part of this Colony.

52. Mezoneuron, a genus similar in habit to the last, has M. *brachycarpum*, from the Richmond River, and M. *Scortechini*, recently described by Baron Mueller.

53. The genus Cassia has about thirty species for all Australia, fifteen of which are found in different parts of New South Wales, but only two in the neighbourhood of Sydney, and one of these is probably not indigenous. The Baron gives the following for New South Wales:—C. *Brewsteri*, C. *lævigata*, C. *sophera*, C. *pleurocarpa*, C. *glauca*, C. *australis*, C. *pruinosa*, C. *circinata*, C. C. *phyllodinea*, C. *eremophila*, C. *artemisioides*, C. *Sturtii*, C. *desolata*, C. *concinna*, and C. *mimosoides*. The flowers of these are for the most part yellow or orange, the stamens usually ten, but some of them frequently reduced to small staminodia, the leaves pinnate with opposite leaflets, and the pods cylindrical or flat.

The leaves of some Australian Cassias are medicinal, and allied
to the true senna, and many of the species of the interior have a
white or hoary appearance.

54. Petalostylis *labichioides* belongs to a genus so called from
the petal-like appearance of its style. It is a shrub with yellow
flowers and pinnate leaves, and occurs in the arid parts of the
interior.

55. Bauhinia *Carronii* has the usual characteristics of the
genus, but its leaflets are quite distinct. The flowers are scarlet
and the pod coriaceous. The specific name was given in honor
of the late Mr. Carron, the botanist in Kennedy's disastrous
expedition in 1848.

SUB-ORDER 3. MIMOSEÆ.

The flowers in this division of Leguminoseæ are small, regular,
and sessile in spikes or heads ; and, in proportion to the number
of species, the genera are few. A good idea of the inflorescence
may be formed from the flowers of the common wattle.

56. Neptunia *gracilis* (sometimes called "sensitive plant," from
the fact that the leaflets close when touched) is an under-shrub,
growing in water and moist places in the hotter districts of the
Colony. The flowers are in small globular heads, leaves bipinnate with numerous leaflets, and pods short, broad, and flat.

57. The genus Acacia is a very large one in Australia, containing upwards of 300 species, and although most marked
as regards the nature of its inflorescence, yet there is a great
difficulty in subdividing the species. This genus includes all the
shrubs or trees called Wattles, Mimosas, Hickory, Sally, Myall,
Brigalow, Mulga, Yarran, &c., and according to the most recent
arrangement the species are grouped in nine sections. In the
first section, or Alatæ, the phyllodia are decurrent, forming two
opposite wings to the stem. The species of this section are
peculiar to Western Australia ; and of the second section, the
Continuæ, which have narrow phyllodia, rigid, tapering into a
pungent point, and continuous with the stem, only one species
(A. *continua*) extends to New South Wales. The third section,
or Pungentes, including rigid shrubs with phyllodia articulated
on the stem and tapering into pungent points, has the following
species in this Colony :—A. *spinescens*, A. *lanigera*, A. *trinervata*,
A. *colletioides*, A. *siculiformis*, A. *tetragonophylla*, A. *juniperina*,
A. *asparagoides*, and A. *diffusa*. Of the fourth section, or
Calamiformes, in which the species have sometimes not any
phyllodia, or else narrow-linear ones, terete and tetragonous, only
four are recorded—A. *rigens*, A. *pugioniformis*, A. *juncifolia*, and
A. *calamifolia;* fourteen are indigenous in Western Australia.
Of the fifth section, or Brunoideæ, only two, A. *Baueri* and A.
conferta, are peculiar to New South Wales and Queensland. These
have phyllodia small and somewhat verticillate. The sixth section,

or Uninerves, comprehends many species, with one central or nearly marginal nerve in the phyllodia, which vary very much in shape, and are occasionally spinescent. The western species are rather more numerous than the following eastern ones, only four being common to both:—A. aspera, A. armata, A. vomeriformis, A. plagiophylla, A. acanthoclada, A. obliqua, A. acinacea, A. lineata, A. hispidula, A. undulifolia, A. flexifolia, A. microcarpa, A. montana, A. verniciflua, A. leprosa, A. stricta, A. sentis, A. fasciculifera, A. falcata, A. penninervis, A. neriifolia, A. pycnantha, A. notabilis, A. gladiiformis, A. obtusata, A. rubida, A. amœna, A. hakeoides, A. salicina, A. suaveolens, A. subulata, A. linifolia, A. crassiuscula, A. buxifolia, A. lunata, A. brachybotrya, A. podalyrifolia, A. vestita, A. cultriformis, A. pravissima, A. myrtifolia. Section 7, the Plurinerves, has the phyllodia vertically flattened with two or more longitudinal nerves. This section includes the species with hard scented wood, and also several trees with timber which can be utilised for carpentry and cabinet purposes—A. pravifolia, A. amblygona, A. elongata, A. subporosa, A. homalophylla, A. pendula, A. Oswaldi, A. stenophylla, A. sclerophylla, A. viscidula, A. ixiophylla, A. melanoxylon, A. implexa, A. venulosa, A. farinosa, A. complanata, A. binervata. Section 8, the Juliflorœ, is chiefly distinguished by its cylindrical and dense spikes of flowers. Of the sixty-nine species, only thirteen occur in New South Wales, and nine in Victoria. Of these, A. glaucescens is a tree of some size, and A. pycnostachya, A. subtilinervis, A. longifolia, A. linearis, A. cyperophylla, A. aneura, A. Kempeana, A. triptera, A. doratoxylon, A. Cunninghamii, A. oxycedrus, and A. verticillata, mostly shrubs. Section 9, the Bipinnatæ, contains one tree rising to 60 feet and upwards, and several species useful for their bark and gums, as well as ornamental in their appearance. This section is well defined by its bipinnate leaves, and one-half of the species extend to Eastern Australia, four being peculiar to New South Wales, A. elata, A. pruinosa, A. spectabilis, A. polybotrya, A. discolor, A. decurrens A. dealbata, A. cardiophylla, A. leptoclada, A. pubescens, A. Farnesiana. The genus Acacia, whether considered in an industrial, medicinal, or ornamental view, is one of the most important and characteristic in Australia.

58. Albizzia (including Pithecolobium and Inga) has for New South Wales A. amœnissima, A. pruinosa, A. Tozeri, and A. Hendersoni. The second of these, which is a beautiful tree, is found as far south as Illawarra. To this genus belongs the celebrated "Rain-tree" of America, A. Saman, which Baron Mueller has recommended for cultivation in this Colony.

From a review of the large order Leguminosæ, it appears that it includes fifty-eight genera and 323 species for New South Wales, the total number for all Australia being 1,058, the largest of the

natural orders for this continent. The species recorded for
Queensland rather exceed those of New South Wales, whilst
those of Victoria are only 168. Of the genera represented in
Queensland, and not in New South Wales, about twenty are
common to India and Australia, thereby giving an Eastern-Asiatic
character to the vegetation of the former colony. Acacia
Farnesiana is a shrub common to tropical parts of the old and
new world, and A. *decurrens*, in one or other of its black or green
varieties known popularly as "wattles," is a most widely-distributed species, extending from Tasmania, through South
Australia, Victoria, and New South Wales, into Queensland.
In the early days of the Colony Callicoma *serratifolia* was called
black-wattle, but now that name is given to the early-flowering
variety of A. *decurrens*. Amongst introduced leguminous plants
the following have established themselves in different parts of
the Colony:—Argyrolobium *Andrewsianum* (Stend.), Medicago
sativa (Linn.), M. *lupulina* (Linn.), M. *maculata* (Willd.), M. *denticulata* (Willd.), Trifolium *repens* (Linn.), T. *pratense* (Linn.),
T. *glomeratum* (Willd.), T. *procumbens* (Linn)., T. *arvense* (Willd.),
Vicia *hirsuta* (Koch.), V. *sativa* (Linn.), Lotus *tetragonolobus*
(Linn.), Ulex *Europæa* (Willd.), Cajanus *bicolor* (Dec.), and
Melilotus *parviflora* (Des.)

2. The natural order ROSACEÆ, though a numerous one in the
more temperate parts of the Northern Hemisphere, is very
limited in Australia, nine species only occurring in New South
Wales. Geum *urbanum*, which is supposed to be indigenous, is
a small shrub with pinnate or pinnatisect leaves, yellow flowers,
numerous stamens, and plumose style. It is common to many
parts of Europe and Asia. The genus Rubus or Brier has the
following species here:—R. *parvifolius*, R. *rosifolius* (a species
widely spread in Africa and Asia), R. *moluccanus*, and R. *Moorei*.
The second is well known in gardens by its double flowers; and
whilst one of the species has edible fruit, the whole are
characterised more or less by astringent properties. Alchemilla
vulgaris has small greenish flowers and leaves palmately lobed.
It is found also in Victoria, and many distant parts of the
world. Agrimonia *Eupatoria* (the common Agrimony of Britain),
a small plant with yellow flowers and pinnate leaves, is supposed
to be indigenous; and Acæna *ovina* and A. *sanguisorbæ*, common
to several of the Australian Colonies, may be distinguished by
their globular heads of small flowers, pinnate leaves, and prickly
fruits. Rosa *rubiginosa* (Linn.), the "Sweet-brier" has become
a very troublesome weed in various parts of New South Wales.

3. The order SAXIFRAGEÆ is nearly allied to the last, and com-
prises some very interesting shrubs and trees. Argophyllum
Lejourdanii extends from Queensland to the Northern districts of
New South Wales. It is an elegant shrub, with panicles of

small flowers, and leaves silvery white on the under surface. Abrophyllum *ornans* is also an ornamental shrub of larger size, with very long leaves, dense panicles of small yellowish flowers, and dark-coloured berries. Cuttsia *viburnea* (Fragmenta, yol v, p. 47) is a shrub of sub-tropical Australia, having panicles of white flowers, leaves similar to those of Abrophyllum, and minute capsules. Colmeiroa *carpodetioides* (Frag., vol. vii, p. 149) is an evergreen shrub from Lord Howe's Island, and added to our Flora since the publication of the second volume of the Flora Aust. Quintinia *Sieberi* is a fine tree, found principally on the Blue Mountains and at Illawarra, flowers white and racemose, leaves coriaceous and reticulate, and capsules opening into teeth or valves. Q. *Verdonii*, which is nearly allied to the preceding, grows on the banks of some of our northern rivers. Polyosma *Cunninghami* seems peculiar to New South Wales. It has opposite leaves, which turn black in drying, racemes of whitish tetrandrous flowers, and ovoid berries. This is a small tree, and the genus derives its name from the scented flowers of the species. Anopterus *Macleayanus* is a handsome shrub, with long narrow leaves, and racemes of white flowers. Callicoma *serratifolia* once occurred frequently where Sydney now stands, and it is still found about the harbour of Port Jackson. The flowers are yellowish, and in dense globular heads, and the leaves opposite, coarsely serrated, coriaceous, shining above and white underneath. Aphanopetalum *resinosum* is an ornamental straggling shrub, with very minute petals, opposite, thick, shining, and serrated leaves, and branches rough with glandular dots. The genus Ceratopetalum appears limited to this Colony. C. *gummiferum* is the well-known Christmas-bush of the colonists, and C. *apetalum*, with the allied Schizomeria *ovata*, furnishes the lightwood and coachwood of workmen. Acrophyllum *venosum* is a very elegant shrub growing on moist shady rocks. It is very rare, and, so far as known, limited to certain parts of the Blue Mountains. The leaves of this plant are principally in threes or ternate, and the flowers pink in dense axillary clusters. Weimannia, a genus of trees or shrubs with opposite digitate leaves and racemose flowers, has four species—W. *paniculosa*, W. *Benthamii*, W. *lachnocarpa*, and W. *rubifolia*. These occur on the Clarence, Tweed, and Hastings rivers, and the second sometimes rises to the height of 100 feet, and is known by the native name "Marrara." Eucryphia *Moorei*, which is found on the ranges to the south, is a handsome tree, with pinnate leaves and large white flowers. There is some doubt as to the proper position of this tree, many botanists placing it with the Hypericineæ on account of its hypogynous stamens and imbricated sepals and petals. Bauera *rubioides* and B. *capitata* are pretty shrubs, with pink or white flowers, and leaves six-whorled, occurring in rocky places near water, or in crevices of rocks on the mountains. The variety *microphylla* is a slender plant, with leaves scarcely two lines long. Davidsonia *pruriens* (a tree figured by F.v.M.) is found also in New South Wales.

4. CRASSULACEÆ.—This order, popularly termed "Houseleeks," though somewhat numerous and extending over the greater part of the world, is represented in New South Wales by four species of Tillæa—T. *verticillaris*, T. *purpurata*, T. *macrantha*, and T. *recurva*—insignificant herbs, some growing in rocky and gravelly, and others in moist places, or even in the water.

5. The order ONAGREÆ is a small one in this Colony, being restricted to one species of Epilobium (E. *tetragonum*), and two of Jussiæa (J. *repens* and J. *suffruticosa*), the former with pink and the latter with yellow flowers. Some authors make several species of Epilobium, but the Baron, having found that they run into each other, has reduced them to one. This plant is used as a rustic remedy in certain urinary disorders. J. *repens* is a creeping plant, sometimes floating in water. The introduced plants of this order are the evening primroses, Œnothera *biennis* (Linn.) and Œ. *rosea* (Willd.)

6. The LYTHRIAREÆ have in the western part of the Colony the small herb Ammania *multiflora* and two species of Lythrum (L. *salicaria* and L. *hyssopifolia*), the one with showy reddish, the other with light bluish flowers. These plants are common to many parts of the world, and are found in creeks and swampy places. In Lythrum the calyx is tubular, and from eight to twelve ribbed.

7. The order HALORAGEÆ consists for the most part of herbs or small shrubs, interesting to the botanist rather than to the casual observer, as they have inconspicuous flowers and occur most frequently in or near water. The genus Haloragis, which is octandrous, has the following species in New South Wales, viz., H. *mucronata*, H. *elata*, H. *ceratophylla*, H. *odontocarpa*, H. *serra*, H. *glauca*, H. *alata*, H. *micrantha*, H. *heterophylla*, H. *pinnatifida*, H. *tetragyna*, H. *teucrioides*, H. *salsoloides*, H. *monosperma*, H. *depressa*. Loudonia *Behrii* is more pretentious, having yellow flowers in dense corymbose panicles, and occurring on sandy soil in the south-western parts of the Colony. Meionectes *Brownii* is a diffuse herb, similar to Haloragis, but it is tetrandrous, and differs from that genus in the binary arrangement of the flowers. Myriophyllum is a genus of aquatic plants common to all parts of the world, and the stems of the species consist of necklace-shaped cellular tissue radiating from the centre. Our species are M. *variifolium*, M. *elatinoides*, M. *verrucosum*, M. *latifolium*, M. *gracile*, and M. *integrifolium*. Ceratophyllum *demersum* is also an aquatic plant, with whorled leaves divided into linear

dichotomous segments, and small axillary flowers, which are unisexual without perianth. Callitriche *verna* and C. *Muelleri* are herbs of similar character, with opposite leaves and fruit, small, four-celled and four-lobed. Both of these genera are common to other parts of the world.

8. Rhizophora *mucronata* and Bruguiera *gymnorrhiza* are the only species of the MANGROVE family which extend to the subtropical parts of Australia. They are smooth evergreen trees, and known by the curious habit of germinating while the seeds are still adhering to the branch that bears the fruit. These, as well as other species of the order, grow near the shore, and root in the mud, forming dense thickets.

9. MYRTACEÆ.—This is the most important order in Australia, whether considered in reference to the value of its timber, the medicinal and industrial properties of its species, the general character of its inflorescence, or the utility of its berried fruits. Whilst in Europe the myrtle family is represented only by M. *communis* (Willd.), a great favourite of the Greeks and Romans, and sacred to the goddess Venus, the bush of New South Wales consists principally of myrtaceous trees and shrubs. This order is well defined by its many-celled ovary, many-petalled, or, in some genera, apetalous flowers, imbricated calyx, numerous stamens, and usually by opposite dotted leaves with a marginal vein. The flowers are red, white, or yellow, and the order naturally divides itself into two sections, viz., that with capsular and that with baccate fruit. According to Baron Mueller's estimate the species of Myrtaceæ rank next to Leguminosæ in point of numbers, being 651 for all Australia, of which 140 are found in New South Wales, whilst in West Australia there are about 380. This great disparity arises from the fact that only three species of Darwinia occur in New South Wales, whilst Verticordia, Pileanthus, Calycothrix, Lhotzkya, Wehlia, Astartea, Hypocalymma, Balaustion, and Agonis have only a few species common to the eastern and western shores. Darwinia, a genus of heath-like shrubs, reckoning more than thirty species in Western Australia, has only D. *taxifolia*, D. *fascicularis*, and D. *virgata* in New South Wales. The flowers are small and pinkish, calyx ribbed, and stamens ten. Homoranthus *virgatus* (Fl. of Australia, vol. iii, p. 15) is reduced to D. *virgata* (Frag., vol. ix, p. 176). Calycothrix has in West Australia twenty-eight species, but only two—C. *tetragona* and C. *longiflora*—common to the Australian Colonies and Tasmania. These interesting plants have the lobes of the calyx produced into hair-like awns. Thryptomene is another genus of heath-like shrubs with white and pink flowers. Our species are T. *miqueliana*, T. *ciliata*, and T. *minutiflora* (including Genetyllis *alpestris* of Lindley, "Mitchell's Expeditions," vol. ii, 178). Bæckia is a large genus in Western Australia, having

there upwards of forty species. Fourteen are found in the eastern part of the continent, only two of which are common to Western Australia. Most of these are very small shrubs, with white flowers and minute strongly-scented leaves. B. *linifolia* and B. *virgata* grow in moist places, or near creeks, and are tall erect shrubs; but the following are not so conspicuous:—B. *diffusa*, B. *crassifolia*, B. *ericæa*, B. *crenulata*, B. *brevifolia*, B. *Gunniana*, B. *diosmifolia*, B. *stenophylla*, B. *camphorata*, B. *Cunninghamii*, B. *densifolia*, and B. *Behrii*.

Leptospermum (the species of which are frequently called tea-trees, or tea-tree scrub) has eleven species, ranging for the most part from Port Jackson to the Blue Mountains, but several occurring beyond the Dividing Range. They are:—L. *lævigatum*, L. *scoparium*, L. *arachnoideum*, L. *lanigerum*, L. *parviflorum*, L. *stellatum*, L. *attenuatum*, L. *myrtifolium*, L. *myrsinoides*, and L. *abnorme*.

It is said that the leaves of L. *scoparium* were used by Captain Cook's ships' crews for the purpose of making a beverage which they called tea, and hence it was that in process of time the name of tea-trees was given to all the species of the genus. The leaves were also used with spruce leaves in equal quantity to correct the astringency of the former in brewing beer from them.

Kunzea is a genus allied to the last, but differing in its exserted stamens. K. *corifolia*, with its white flowers, and K. *capitata*, with its purple ones, are shrubs occurring near Sydney, whilst K. *parvifolia*, K. *Muelleri*, and K. *peduncularis* are southern species.

Callistemon, so called from the showy red stamens of some species, seems to pass into Melaleuca; and the species are so similar in their floral characters that, without reference to their foliage, they might be regarded as varieties of one form. They are C. *lanceolatus*, C. *coccineus*, C. *salignus*, C. *rigidus*, C. *linearis*, C. *pinifolius*, C. *pithyoides*, and C. *brachyandrus*. The colour of the flowers varies from green and pale yellow to crimson, the stamens being sometimes an inch long. One species, called "broad-leaved tea-tree," is a tree of some size, and has very hard wood.

Melaleuca is a large genus, of which by far the greater number of species, that is seventy-two, are endemic in Western Australia. About eighteen belong to New South Wales, and these have white, purple, or crimson flowers. M. *linariifolia*, M. *Leucadendra*, M. *genistifolia*, and M. *styphelioides* are trees of some size, with white flowers; and from the leaves of the second cajeput oil can be distilled. A decoction of the leaves is much used in China as a tonic, and the bark can be utilised for caulking boats and covering rude buildings. M. *hypericifolia* is a fine shrub with showy crimson flowers. M. *thymifolia* and M. *erubescens* are much smaller plants, with purple or reddish flowers, whilst the remaining species are somewhat similar in character, but varying in size, and having white flowers. These are—M. *acuminata,* M. *pauciflora,* M. *squarrosa,* M. *parviflora,* M. *armillaris,* M. *uncinata,* M. *hakeoides,* M. *squamea,* M. *nodosa,* M. *ericifolia,* and M. *pustulata.*

Angophora is a genus of which strangely enough the species are called " Apple-trees," and of the four only one extends to Victoria and three to Queensland. Though nearly allied to Eucalyptus, the genus differs in having distinct petals, prominent calyx teeth, and capsules opening sideways in three or four valves. The leaves, too, are opposite. A. *cordifolia,* the smallest of the apples, is a tall shrub, growing near the coast, with bristly or reddish hairs on the branchlets and inflorescence, largish flowers, and fruit sometimes nearly an inch in diameter. A. *subvelutina* is a large tree with persistent fibrous bark, sessile leaves, and small flowers in loose corymbs. It frequently gives the name of " apple-tree flats" to places where it occurs. A. *intermedia* differs in being glabrous, and in having the leaves distinctly petiolate. The variety called "narrow-leaved apple" flourishes in sandy or rocky places. A. *lanceolata* (the " Red-gum" of workmen near Sydney) is a large tree, and when the bark has fallen off it becomes smooth and similar in appearance to the spotted-gum (E. *maculata*). Two of the trees popularly called " apples" afford timber for the naves of wheels and rough carpentry, and the so-called " Red-gum" is abundant in a red-coloured resin and useful for fuel. Sir Wm. Macarthur looked upon this tree as a connecting link between Angophora and the smooth-barked Eucalypts.

EUCALYPTUS.

The genus Eucalyptus is one of great difficulty to botanists; and the description and grouping of the species, though considerably assisted by the labours of Mr. Bentham and Baron Mueller (especially of the latter, in his " Eucalyptographia"), may yet be regarded as only partially accomplished. With respect to the description of species, it may be observed that trees differing widely in the nature of their wood and the texture of their bark have very similar flowers, so that in determining simply from dried specimens there is a danger of associating together very different species. A knowledge, therefore, of the tree in its living state—and that, too, in its various stages of growth—as well as a due consideration of the character and extent of its bark, seems necessary for fixing the species with any degree of certainty. And then when species have been described (of which probably there may be 150), the other great difficulty presents itself in the

grouping of them, or placing them in separate sections. In the
early days of the Colony, when only a few species were known, it
was considered quite sufficient to classify them according to the
comparative length of their operculum, or the lid which covers
the stamens in the bud ; but when it was found that this arrangement caused the separation of species closely allied, and, further,
that in the same species the operculum was not always of the
same shape and length, Baron Mueller devised a method which,
for practical men, is of great utility, viz., that of grouping species
according to the nature of their bark ; that is to say, placing the
smooth-barked, half-barked, fibrous-barked, and deeply-furrowed
barked trees in their respective sections. This arrangement (though
some trees vacillate between fibrous and smooth, and others verge
towards different sections, according to their age) is certainly one
of the most natural which has been devised, and has had a very
good effect in correcting mistakes. It has not found favour,
however, with European botanists. Not having much knowledge
of the trees in a living state, and being unable to appreciate the
utility of a system which seems to ignore the characteristics by
which species are generally grouped, Mr. Bentham conceived
that a mode of classification founded on the shape and opening
of the anthers might be highly useful for associating allied species
in herbaria, as well as for distinguishing them from each other by
a well-defined character. This system is very ingenious, and, like
Baron Mueller's cortical system, it has helped to clear up some
of the errors into which the early botanists had fallen ; but it
cannot be pronounced altogether satisfactory, because it is too
artificial in its character, and labours under the defect of separ-
ating species which in the minds of the colonists must ever be
placed together. Thus, for instance, those trees which are
regarded as allied to each other, viz., the various species of
" mahogany," so called, and also of ironbark, which, one would
think, should stand in close proximity to each other, are placed in
different sections. Baron Mueller has followed this classification,
with certain modifications, in his " Census of Australian Plants,"
but he has stated his intention of paying particular attention to
the shape and organic structure of the fruit in the respective
species with a view of founding a system on purely carpologic considerations. Until, therefore, some more satisfactory grouping
of species is adopted by botanists it may be advisable to follow
the Baron in dividing them into—(1) Renantheræ, or such as
have kidney-shaped anthers ; (2) Porantheræ, those whose
anthers open in small circular pores; and (3) Parallelantheræ,
the largest of the divisions, such as have the anthers ovate or
oblong and opening in parallel slits.

(1) RENANTHERÆ.

1. E. *stellulata* is a tree of moderate size, found in the mountainous parts of New South Wales, having for the most part smooth bark of a white or lead colour, but occasionally with fibrous bark on the butt. It is remarkable for the almost longitudinal veins of the leaves and its minute starlike buds. In different parts it is called white, green, or lead-coloured gum, sallee, &c. The shrubby form about the higher parts of the Blue Mountains is now regarded as a variety under the name E. *microphylla*. It has narrow leaves, and fibrous bark on the lower part of the butt.

2. E. *pauciflora* (Cunningham's E. *coriacea*) is also a tree of moderate size, known as one of the "white gums," and limited to hilly and mountainous regions in the southern parts of the colony and New England. The leaves are of a leathery texture, and their veins rise nearly together from the base.

3. E. *regnans* a very large tree with a smooth bark, and found only in the Southern forests. In Victoria it attains a height of more than 400 feet, and represents the loftiest tree in British territory. Baron Mueller, who formerly regarded this as a variety of E. *amygdalina*, says that in the Dandenong Ranges it is called " White-gum " and " Mountain ash."

4. E. *amygdalina* is generally known by the name of " Messmate," and on the Mittagong Range and elsewhere rises to the height of 150 feet and more. In its typical form it is reckoned as one of the half-barked trees, the upper branches being smooth, but the bark varies in different places; and if Baron Mueller's opinion be correct, E. *radiata* (a white-gum growing on the banks of the Nepean and elsewhere) must be regarded as a variety of E. *amygdalina*. There is so great a similarity in the flowers, fruit, and leaves that it seems almost impossible to avoid the conclusion, though the trees differ very much in bark, wood, and habit.

5. E. *obliqua* is remarkable as having been the first species of eucalypt known to Europeans (F. v. M.), and in South Australia and Tasmania it is called " Stringybark," whilst in Victoria it is regarded as the "messmate." Baron Mueller considers that in its truly typical form it scarcely reaches the southern limits of New South Wales. This species differs from the other stringybarks in having the capsule more or less sunk.

6. E. *stricta* is a shrubby species common on the elevated parts of the Blue Mountains, having long narrow leaves similar to those of E. *microphylla*, but differently veined. This appears to be the mountain variety of E. *obtusiflora*, the only shrubby form found near the coast, but with larger flowers and leaves. Both of these eucalypts have smooth bark, similar anthers, and depressed capsules.

D

7. E. *macrorhyncha* is the form of Stringybark most common in elevated situations. It occurs on the Southern ranges, the hills beyond Mudgee, and also in New England. Baron Mueller notes as peculiar to this stringybark "fruit calyx almost hemispherical, not much longer than the amply protruding very convex vertex ; valves wholly exserted."

8. E. *capitella* is the form of Stringybark which occurs most frequently near the coast, and it is one of those species described in Willdenow, 1799. It is nearly allied to the preceding species and E. *eugenioides*, but the umbels of flowers are capitate.

9. E. *eugenioides* occurs in forest lands from Port Jackson to the Blue Mountains. It approaches very near to E. *capitella* and perhaps is only a variety of it. The chief difference is that it has narrower leaves, shining on both sides, and "calyxes often attenuated into short stalklets" (F. v. M.)

10. E. *piperita*, or the Peppermint (Willd., 1799), was well known to the early colonists, as it occurred probably where Sydney now stands. It derived its specific name from the volatile oil distilled from its leaves. The fruit of this tree is more or less truncate-ovate, and the bark, which is less fibrous than stringybark, does not extend to the upper branches.

11. E. *pilularis*, or Blackbutt (the specific name of which is not well chosen, seeing that other species have fruits more pillular in shape), is a half-barked tree, rising to the height of 150 feet and upwards. This species prefers a good soil, and is of rapid growth. The young saplings have opposite leaves, sessile, and of a lanceolate shape.

12. E. *acmenioides*, or the White Mahogany, though placed as a variety of the preceding in the "Flora Australiensis," is certainly a distinct species, differing from it in having fibrous persistent bark, similar to that of the stringybark, smaller flowers and fruits, and leaves not so oblique. It used to be plentiful in the woods to the north of Parramatta, and in other parts near the Eastern coast. Baron Mueller identifies this tree as E. *triantha* (Link.), but he considers the name inappropriate.

13. E. *hæmastoma*, or "White-gum," was described also in 1799, and it derived its specific name from the red rim of the fruit. Though usually placed amongst smooth-barked trees, it has a variety with fibrous bark on the butt. The form E. *micrantha*, with much smaller flowers and fruits, occurs here and there in poor soil from Parramatta to the Blue Mountains.

14. E. *Sieberiana* (the E. *virgata* of some botanists) is the tree known as Mountain-ash, Gum-top, or even Ironbark. It is a half-barked tree, the bark on the butt being dark coloured and deeply furrowed like Ironbark, whilst the branches are smooth and sometimes white. E. *Siberiana* may be regarded as an alpine species.

15. E. *microcorys*, called Wangee or Forest Mahogany, is a
large tree "dispersed from the vicinity of Cleveland Bay to the
Hastings River" (F. v. M.) The bark is persistent, lamellar,
and much less fibrous than that of other so-called mahoganies.
This tree furnishes the tallow-wood of commerce.

16. E. *dives*, or the broad-leaved Peppermint, is a small tree
occurring in forests beyond the Blue Mountains, and also on the
Mittagong Range. The lower leaves are opposite, cordate and
acuminate; the upper ones like those of E. *obliqua*, as well as
the buds and anthers, but the fruit is nearly hemispherical,
with a thick-furrowed rim, and valves not protruding. Mr.
Bentham was inclined to regard this as a variety of E. *obliqua*,
which it also resembles in bark, but he had not seen the fruit.
(E. *Planchoniana* and E. *Baileyana* have recently been found in
the Northern districts.)

(2) PORANTHERÆ.

17. E. *paniculata* is the hardest of all the Ironbarks, being
known amongst workmen as the "White Ironbark," because
the wood is paler than that of its congeners. Though on this
side of the Dividing Range it attains the height of 100 feet and
more, yet in the interior it assumes a stunted or shrubby appearance. There is a narrow-leaved variety of this species very
similar to E. *crebra*, and it can scarcely be distinguished but by
the opening of the anthers.

18. E. *sideroxylon* is an Ironbark frequently with red flowers,
deeply-fissured bark, and darker wood than any of the kind. It
occurs between Parramatta and Liverpool, in the neighbourhood
of Richmond, and also in several parts beyond the Dividing Range.
There is a tree in South Australia and Victoria very similar to it
in flowers and fruit, but differing from it in the nature of the
bark and colour of the wood. This is E. *leucoxylon*, of which,
according to the anthereal system, E. *sideroxylon* is regarded as
a variety.

19. E. *melliodora*, or the "Yellow Box," is a tree of moderate
size, found near Bathurst and Mudgee, as well as in New England and on the Darling. It is placed amongst the half-barked
trees, because the bark, though for the most part persistent on
the stem, scales off from the branches. The inner bark is of a
yellowish colour, and the flowers have the scent of honey.

20. E. *polyanthema*, sometimes called "Lignum Vitæ or Bastard
Box," is a tree of moderate size, with a rough bark persistent on
the stem and branches, and leaves broadly ovate-orbicular, frequently acuminate. This species, which is remarkable for the
toughness of its wood, grows on the banks of creeks in the coast
districts, and only occasionally as a forest tree. It is found
beyond the Dividing Range with bark similar to that of "Box,"
and leaves smaller and not acuminate.

21. E. *populifolia*, the "Shining Poplar-leaved Box," has sometimes been mistaken for the preceding species, but it occurs only
in the warmer parts of the Colony, and differs from E. *polyanthema*
in having all its stamens fertile, and its anthers with more lateral
openings; the filaments likewise are usually darker and the fruit
smaller. A small tree called "Slaty Gum," and growing on
ridges in the western interior, resembles this species in flowers
and fruit, but it has paler leaves and smooth bark. According to
the anthereal system, it bears the same relation to E. *populifolia*
that E. *radiata* does to E. *amygdalina*.

22. E. *ochrophloia*, or the "Yellow Jacket" of the Warrego
and Paroo, has smooth bark, and rises occasionally to the height
of 50 feet, though generally much less. The leaves are from 4
to 6 inches long, of a falcate or oblong-lanceolate shape, and the
fruit nearly half an inch long, the valves short and enclosed. (See
Fragmenta, vol. xi, p. 36.)

23. E. *gracilis* is a shrubby species from the Mallee-country
on the Murray and Darling, usually only a few feet in height,
and having several stems from one root. According to Baron
Mueller, it forms with E. *incrassata*, E. *uncinata*, E. *oleosa*, and
E. *paniculata*, the extensive Mallee scrubs of extra-tropical
Australia. The leaves are small and narrow, the anthers very
minute, and the fruit somewhat urn-shaped and angular.

24. E. *uncinata* is similar to the last in size and habit, but it
has filaments bent inwards even when fully expanded, and fruit
semi-ovate in shape. It differs also from E. *gracilis* in having all
the anthers fertile, and the valves of the fruit slightly exserted.

25. E. *largiflorens*, in the coast districts, is a large half-barked
tree, growing in moist places, and generally known as "Bastard
Box." On the river-flats of the interior it is rather diminutive,
and has longer and narrower leaves, crimson filaments, and a
drooping habit. This species is the same as E. *pendula* and
Cunningham's E. *bicolor*, probably so called on account of the
flowers varying from cream-colour to crimson.

26. E. *Behriana*, though more frequent in South Australia
and Victoria, extends to the Lower Darling. It resembles the
box in its leaves and flowers, but differs from that species in its
shrubby habit and smoother bark, whilst the flowers and fruit
are much smaller. Like the preceding, the filaments are sometimes crimson. According to Baron Mueller, "the resemblance
of E. *Behriana* in foliage is closer to E. *hemiphloia*, but in flowers
and fruits nearer to E. *largiflorens*."

27. E. *hemiphloia* (including E. *albens* or the "White Box" of
the interior) is the common "Box" on this side of the Dividing
Range, and forms a large part of the forest trees, indicating

generally good grazing country, and rising to 100 feet and more in height. In this species the anthers are very minute, the leaves large, and the valves of the fruit enclosed.

28. E. *melanophloia* is the " Silver-leaved Ironbark," common to Queensland and the northern parts of New South Wales. When Mitchell was on the Narran, in 1846, he regarded this tree as a novelty, the leaves of it being different from those of other Ironbarks in the Colony. This species does not attain any great size, and the leaves are sessile and opposite, the wood also being inferior to that of other Ironbarks.

(3) PARALLELANTHERÆ.

29. E. *crebra* is usually called the " Narrow-leaved Ironbark," and occurs from Port Jackson to the Blue Mountains. It prefers ridges and ranges, and also a better soil than some of the other Ironbarks. The wood is red and durable, though not so strong as that of E. *siderophloia* or E. *paniculata.* The flowers, fruit, and leaves of this species are sometimes very small.

30. E. *siderophloia* is the " Broad-leaved or Red Ironbark," and is justly regarded as one of our most valuable timbers for piles, railway-sleepers, spokes, &c. The Botany Bay Kino was procured principally from this species, and hence Allan Cunningham and other botanists were accustomed to call it E. *resinifera,* a designation now applied to another species.

31. E. *microtheca* is for the most part a half-barked tree of moderate size, common to the north of New South Wales, the arid regions near the Darling, and also to North, South, and Western Australia. According to Baron Mueller's " Select Plants," p. 124, " The wood is brown, sometimes very dark, hard, heavy, and elastic ; prettily marked; thus used for cabinet-work, but more particularly for piles, bridges, and railway-sleepers."

32. E. *globulus,* the "Tasmanian Blue-gum," is a tree of world-wide celebrity, not limited to Tasmania, but extending to Victoria and the Southern districts of New South Wales. Baron Mueller transmitted seeds of this species, in 1853, to several botanic gardens in Europe, and since that period it has been successfully cultivated in the South of Europe, the Northern coast of Africa. the Western States of America, the North of India, &c.

33. E. *longifolia,* or the " Woolly Butt," is distinguished from other eastern species by the arrangement of its flowers in threes, by the length of its leaves (especially in young trees growing near water), and by the large size of its fruit. In the younger trees the bark is persistent on the stem and branches, but in old trees the upper branches, and sometimes even the butt, become nearly smooth.

34. E. *robusta*, the "Swamp Mahogany" of workmen, is a large and handsome tree, admired for its spreading limbs and noble foliage. It is fond of moist and swampy localities, and rises to the height of 100 feet and upwards. Though popularly termed Mahogany, this tree is not in any way allied to the Mahogany of the West Indies, and the wood is not considered durable. The flowers are large, and the fruit more than half an inch in length.

35. E. *botryoides*, sometimes called "Bastard Mahogany" or "Bangalay," is a tree seldom attaining any great size, and having frequently a gnarled appearance. It occurs in sandy places near the coast, and is not found to the west of the Dividing Range.

36. E. *goniocalyx* is the "Spotted-gum" of Victoria, but it extends into New South Wales as far as Braidwood. It derives its specific name from the prominent angles of the calyx. In the rich forest-valleys of Victoria this tree is said to attain a great height, and Baron Mueller regards it as one of the Eucalypts most deserving of cultivation. A stunted variety of this, with fibrous bark on the butt, occurs on the ridges beyond Mudgee.

37. E. *incrassata*, and its small-flowered variety E. *dumosa*, constitute a large portion of what is called the Mallee scrub, and " plays an important part in the natural economy of the desert, aiding to mitigate the excessive heat and the effect of sirocco-like blasts of widely arid regions by its enormous power of evaporation, and also that of its roots in drawing up and absorbing humidity from the soil" (F. von Mueller). The Baron likewise remarks that this is one of the Eucalypts which, like the following species, yields from its roots a supply of water to the parched traveller.

38. E. *oleosa* is another of the species forming a part of the Mallee scrub, and seldom attaining any size. It resembles E. *dumosa;* but in that species the "flower-stalks are generally much dilated, the calyxes, including the lid, are streaked, the anthers are mostly longer in proportion to their width, and the fruit-valves are terminated only in short points" (F. v. M.)

39. E. *Gunnii*, the "Swamp-gum or White-gum" of the Southern districts of the Colony, is not confined to low swampy places, but occurs also on the mountain ranges. It is generally a smooth-barked tree, having top-shaped or bell-shaped fruit, with short valves fixed close to the orifice. In alpine regions it is dwarfed, and flowers when only a few feet high.

40. E. *resinifera,* or " the Red Mahogany," is a lofty tree with a rough persistent bark, and though differing very much in the size of its flowers and fruit, its bark, wood, and exserted valves render it easy of recognition. The larger forms, known as *grandiflora* and *pellita*, though regarded by some as distinct, seem merely a more luxuriant growth, occasioned by additional moisture and proximity to the coast. E. *resinifera* does not extend beyond the Blue Mountains.

41. E. *saligna*.—The specific name of this tree is rather inappropriate for our "Blue or Flooded Gum," as its appearance is not willow-like. This species is found on the banks of creeks, in gullies, or in good alluvial soil, and is one of the finest gums on this side of the range, being rapid in growth, lofty in stature, and sometimes ornamental in its habit.

42. E. *punctata*, the "Leather-jacket or Hickory," is mentioned in the "Flora Australiensis" as a variety of the Grey-gum (E. *tereticornis*), but Baron Mueller has properly described it as a distinct species. The bark, though rougher than that of the ordinary gums, is not persistent, except occasionally on the butt. When it sheds its bark the tree has a reddish or yellowish appearance. Melitose manna is sometimes found on the leaves, and the wood is very tough and durable.

43. E. *pulverulenta* (including E. *cinerea*) is a small tree, very like the apple (Angophora) in its bark, and hence sometimes called "Argyle Apple." The leaves are all sessile, opposite, generally glaucous, and cordate in shape. Martin, in his "History of Australasia," remarks in reference to this tree : "The geology and natural vegetation of a country are intimately connected. In New South Wales the rock which forms the basis of the country may be inferred from the kind of tree or herbage that flourishes on the soil above. For instance, E. *pulverulenta*, a dwarfish tree, with glaucous-coloured leaves, growing mostly in scrub, indicates the sandstone formation ; whilst those open, grassy, and park-like tracts, affording good pasturage, and thinly interspersed with E. *mannifera* (*viminalis*), characterise the secondary ranges of granite and porphyry." This remark applies to the Southern districts.

44. E. *Stuartiana* has a wide range in New South Wales, being found on the Mittagong Range, the hills near Mudgee, and parts of New England. The bark is fibrous and persistent, and it is known by the popular names "Camden Woolly Butt," "Pepper-mint," or "Stringy-bark." On young trees the leaves are frequently opposite. It occurs on the Mittagong Range in company with E. *amygdalina*, and rises to the height of 100 feet.

45. E. *viminalis*, the "Manna-gum" of the interior and a grey gum of the coast districts, is a tree varying considerably in the bark, the shape of the operculum, and even in habit. In the smaller form the leaves have a drooping or willow-like look, and the bark is more persistent and of a darker colour than that of the large trees growing on river-flats in the interior. Baron Mueller regards E. *dealbata* as a variety of E. *viminalis*, but it seems to differ in its glaucous appearance, in the leaves being ovate-lanceolate with more divergent veins, the intramarginal one farther from the edge, and in the fruit being smaller without any raised marginal rim round the orifice. This is a small tree growing on ridges and flowering when young.

46. E. *rostrata,* "the River or Red Gum," beyond the Dividing Range, has a very wide range, being found on the banks of the Cudgegong, Castlereagh, Darling, &c., and is highly valued for its industrial and medicinal properties. It frequently attains a great size, and when growing luxuriantly by the side of the rivers it has often been an object of admiration to explorers. A tree of this species, under which Hume and Hovell are said to have camped in 1824, may still be seen near their obelisk at Albury.

47. E. *tereticornis* is one of our commonest gums, and in different districts is known as "Grey, Red, or Blue Gum" and "Bastard Box." Though distinguished by its bark, wood, and broad and very prominent rim of the fruit, the species is subject to great variation in the length of the operculum, the shape of the leaves, and the size of the capsule, whilst, in moist places, the umbel often exceeds the typical number of flowers. According to the anthereal system, it is allied to E. *rostrata,* but the two trees differ altogether in habit, the one being a forest tree, and the other being truly a river-gum.

48. E. *corymbosa,* or "the Bloodwood," appears as a stunted tree near the coast, but between Sydney and Parramatta it has often exceeded 100 feet. The umbels of flowers are larger than those of many gums, and they are arranged in a somewhat corymbose or paniculated manner. The fruit is urn-shaped, the valves of the capsule deeply enclosed, and the seeds winged. Its specific name is derived from the quantity of red-coloured resin contained in the concentric circles of the wood.

49. E. *terminalis* is regarded by some botanists as a variety of the preceding species, but Baron Mueller, in his recent Census, places them separately. He says that, though in many respects similar, "the fertile seeds of E. *terminalis* are as a rule provided with a terminal membranous appendage of about the length of the kernel, a characteristic hardly ever occurring in the typical E. *corymbosa.*" This eucalypt occurs on the Paroo.

50. E. *eximia,* or "the Mountain or Yellow Bloodwood," differs very much from the other bloodwood. It has large cream-coloured flowers, urn-shaped fruit (sometimes an inch long), and scaly bark, yellowish in colour, persistent on the butt and smooth on the smaller branches. In the month of October this tree may be seen flowering abundantly near the railway line beyond Emu, and then forming a conspicuous object amidst the dark foliage of the forest.

51. E. *maculata,* or the "Spotted-gum," so called from the mottled appearance of its bark, is a fine tree, rising to more than 100 feet, and sometimes 80 or 90 feet without a branch. It has a double operculum like the preceding species, and, according to the systematic arrangement, is nearly allied to it. E. *maculata,*

however, is a smooth-barked tree, and, in addition to a dissimi-
larity of habit, it differs from the mountain bloodwood in the
venation of the leaves and the mode of inflorescence, whilst the
wood, which is of great strength and elasticity, is much more
highly valued.

52. E. *tessalaris*, or "the Moreton Bay Ash," a species extending to New Guinea, is found in the north-east part of New South
Wales.

From a review of the genus, it appears that about one-third of
the known species in Australia are indigenous in New South
Wales, and that the species near the coast are more numerous
than those inland ; whilst, with some few exceptions, the eucalypts
of the interior are either stunted forms of species on the eastern
side of the Dividing Range, or are species of a shrubby kind
peculiar to arid regions.   It is remarkable that only one shrubby
species occurs near Port Jackson, and that that in a smaller form
appears again on the higher parts of the Blue Mountains. Those
species which attain the greatest height are in the Southern
ranges, though none of them are so lofty as the gigantic eucalypts
of Victoria and Western Australia.

Tristania, including Lophostemon, has four species in New South
Wales. T. *neriifolia*, or " Water-gum," a small tree growing on
the banks of creeks, is remarkable for the toughness and elasticity
of its wood, being used for mallets and cogs; T. *suaveolens*, a
shrub somewhat similar in appearance, but differing in habit and
having opposite leaves; T. *conferta*, sometimes called " Red Box "
and " Mahogany," an ornamental tree, and prized for the durability of its timber ; and T. *laurina*, for the most part shrubby,
but in some localities attaining a considerable height.

Metrosideros (with which Syncarpia is now joined) is a genus
widely distributed in the Pacific and Indian Oceans. M. *glomulifera*
(S. *laurifolia*, Benth.), or the " Turpentine," extending from the
Northern districts to Illawarra, is a fine tree, rising to 100 feet
and more, and its wood is said to be proof against the teredo
navalis. M. *leptopelata* is a smaller tree, and found further
inland. M. *polymorpha* and M. *nervulosa* are species indigenous
in Lord Howe's Island. They are described in the " Fragmenta,"
vol. viii, pp. 14 and 15.

Backhousia, commonly known as " Myrtle," is represented
in New South Wales by B. *myrtifolia* and B. *sciadophora*, the
former being widely distributed, and the latter limited to the
Northern districts. B. *myrtifolia* sometimes attains a diameter of
18 inches. The wood is closely grained and prettily marked.

Rhodomyrtus *psidioides* is common to New South Wales and
Queensland. It is a tree of large size, with a diameter exceeding
20 inches and a height of about 40 feet. The fruit is a berry of

an ovoid-globular form, and the wood is reported to be closely grained. This tree is known as far south as Hunter's River.

Of the genus Myrtus, some of the species are shrubs, and others trees varying in height. M. *rhytisperma*, M. *tenuifolia*, and M. *Becklerii* are shrubs, but M. *acmenoides* and M. *fragrantissima* are trees, the former attaining the height of 30 or 40 feet, but not much valued for its timber.

Rhodamnia *trinervia* is a pretty shrub, extending from Queensland to Illawarra. R. *argentea* is a tall tree, not reaching so far south, and distinguished by its silvery-white appearance. The wood is tough and firm.

Eugenia has twenty known species in Australia, seven of which occur in different parts of the Colony. They are for the most part ornamental trees, and several of them have edible fruits which can be utilised for jam. E. *Smithii* ("lillypilly") and E. *myrtifolia* (the native myrtle) are common in the county of Cumberland; but E. *hemilampra*, E. *Moorei*, E. *corynantha*, E. *Hodgkinsoniana*, and E. *Ventenatii* belong to the Northern districts. E. *Moorei* has a good-sized fruit, and E. *Ventenatii* and E. *Smithii* are valued for their wood.

Acicalyptus *Fullageri*, from Lord Howe's Island, is a tree sometimes 120 feet high, and called "Scalybark-tree." It was discovered by Mr. C. Moore, F.L.S., and a full description of it may be seen in the "Fragmenta," vol. viii, p. 15. In Mr. Duff's report of the vegetation of Lord Howe's Island two myrtaceous trees are mentioned under the names of scalybark and honeysuckle, but it appears that the latter name is applied by the islanders to four different trees.

From the Myrtaceæ now enumerated it appears that of that order in New South Wales there are eighteen genera and 149 species. In Western Australia the species are much more numerous, being more than 370, whilst the genera are twenty-five. Angophora, Tristania, Metrosideros, Backhousia, Osbornia, Rhodomyrtus, Myrtus, Rhodamnia, Fenzlia, Decaspermum, Eugenia, Acicalyptus, Barringtonia, Careya, and Sonneratia do not occur in West Australia, whilst the described species of Eucalyptus are forty-six. Of Bæckia and Melaleuca, however, there are 110 species, and of Darwinia and Verticordia, which are almost limited to that colony, nearly seventy species.

In Queensland the species of the order are 132, and in Victoria seventy-eight, the eucalypts of the latter being proportionately large, viz., thirty-five. It is remarkable that of the second section of Myrtaceæ (those which have a berry or drupe) no species are found in West Australia; but that in Queensland, New South Wales, and Victoria they occur in the following order: Queensland, 42; New South Wales, 16; Victoria, 1. Mr. Bentham, after having referred to Eugenia as a genus spread over the tropical and sub-tropical regions both of the old and new world, says that of sixteen Australian species twelve or thirteen are endemic, and three, or perhaps four, common to East India and the Archipelago. Amongst these, E. *Moorei* or E. *jambolana* is common to parts of Queensland and New South Wales. Sir J. D. Hooker, in his introductory essay to the flora of Tasmania, mentions it as a remarkable fact that, whilst a large number of flowering plants characteristic of the Indian peninsula appear in Australia (especially the tropical part of it), not one characteristic Australian genus has ever been found in the peninsula of India.

10. The MELASTOMACEÆ, though numerous in India or the Oriental Archipelago, are limited in Australia to a few species, only one of which, Melastoma *malabathricum*, extends to the northern parts of New South Wales. This plant, common to India and Australia, is a shrub of a few feet in height, with large purple flowers, three or five nerved leaves, and elongated anthers opening at the top in a single pore.

11. RHAMNACEÆ.—This order is characterised by flowers with a fleshy disc, and stamens opposite the petals. Ventilago *viminalis* is a small tree of the west and north-west interior, with long narrow leaves, remarkable for their very oblique and almost parallel venation. The flowers are small and clustered. Alphitonia *excelsa* is a tree of some size, with hard closely-grained wood, which is durable and takes a high polish. In some places it goes by the name of "Red Ash." The genus Pomaderris has in New South Wales the following species:—P. *lanigera*, P. *elliptica*, P. *phillyroides*, P. *vacciniifolia*, P. *Calvertiana*, P. *ledifolia*, P. *apetala*, P. *cinerea*, P. *prunifolia*, P. *ligustrina*, P. *betulina*, and P. *racemosa*. Some of these have flowers in dense corymbs or panicles, and in a shrubbery they are rather ornamental. Emmenospermum *alphitonioides* is a tall tree with persistent seeds similar to those of Alphitonia, but the petals of the flowers are hood-shaped, inserted with the stamens on the margin of the disc. Cryptanda is a large genus, containing more than fifty species, eleven of which occur in different parts of the Colony, and are for the most part heath-like shrubs. The petals are hood-shaped, enclosing the anthers, and inserted with the stamens at the top of the calyx-tube. C. *ericifolia*, C. *propinqua*, C. *spinescens*, and C. *amara* may be found between Port Jackson and the Blue Mountains; but the other eight species farther inland—C. *lanosiflora*, C. *tomentosa*, C. *buxifolia*, C. *longistaminea*, C. *Hookeri*, C. *subochreata*, C. *vexillifera*, and C. *Scortechinii*. Colletia *pubescens* (Discaris *Australis* of Hooker) is a scrubby, thorny shrub, with small opposite leaves and solitary flowers, appearing sometimes quite leafless.

12. The order of the ARALIACEÆ is closely allied to that of the Umbellifers, differing principally in having its ovary more than two-celled, and in its greater tendency to form a woody stem; but Mr. Bentham observes that, though some botanists would unite the two orders, there is really very little difficulty in drawing the line of demarcation between them. Astrotricha is a genus of shrubs, clothed more or less with stellate tomentum, and having large terminal panicles of small flowers articulate on the pedicel. The species A. *longifolia*, A. *ledifolia*, and A. *floccosa* are common in many parts of New South Wales. Panax is represented by five species—P. *Murrayi*, P. *sambucifolius*, P. *cephalobotrya*, P. *elegans*, and P. *cissodendron*. P. *sambucifolius* is a common species resembling the elder in foliage, but P. *elegans* and P. *Murrayi* are fine palm-like trees, rising to the height of 50 or 60 feet without a branch, and throwing out pinnate fronds and large panicles of flowers from their tops. The latter of these occurs on the Blue Mountains, but it is becoming rare. Meryta (Botryodendron) *latifolia* and M. *augustifolia* are enumerated by Baron Mueller amongst the plants common to Norfolk Island (Frag., vol. ix, p. 169) and Tahiti (Endlicher). They are similar in habit to the loftier species of Panax, with the leaves alternately approximate on the tops of the branches.

13. The UMBELLIFERÆ are either herbaceous or shrubby, and in Australia the genera are well marked, only two or three being uncertain in their connection with northern ones. Hydrocotyle consists of prostrate weedy species with small flowers in dense umbels. H. *vulgaris* and H. *asiatica* are common to many parts of the world. H. *laxiflora* is known by its unpleasant scent. H. *geraniifolia* is interesting for its distinctive foliage. The remaining species are H. *hirta*, H. *rhombifolia*, H. *pedicellosa*, H. *tripartita*, H. *callicarpa*, H. *trachycarpa*, and H. *capillaris*. Didiscus (including the species referred to Trachymene in the Flora Aust.) has D. *pusillus*, D. *cyanopetalus*, D. *pilosus*, D. *glaucifolius*, D. *eriocarpus*, D. *procumbens*, D. *incisus*, and D. *humilis*. Some of these plants have edible roots, and are popularly termed "native carrots." Trachymene (the Siebera of the Flora Aust.) is more shrubby than the preceding, and of the fifteen species common to Australia the following are frequent in this Colony:—T. *ericoides*, T. *linearis*, T. *Stephensonii*, and T. *Billardierii*. The last of these varies considerably in the shape of the leaves. All the species have small white flowers and the fruit laterally compressed. Xanthosia (of which five species, X. *tridentata*, X. *pilosa*, X. *dissecta*, X. *vestita*, and X. *Atkinsonia*. are distributed through different parts of New South Wales) is a genus of shrubby plants, diffuse or straggling in habit, and frequently clothed with soft hairs and stellate tomentum. The flowers are yellowish, usually in compound umbels and enclosed in bracts. Actinotus *Helianthi* and A. *minor*

are common in the neighbourhood of Sydney, and have the appearance of composites rather than of Umbellifers. The flowers are surrounded by a radiating involucre, and are often tomentose or woolly. A. *Gibbonsii* is a species from New England described in the Fragmenta, vol. vi, p. 23. Eryngium is represented as a genus with prickly leaves and involucres. The flowers are mostly blue and in heads or spikes, the bracts being sometimes very rigid. E. *rostratrum*, E. *vesiculosum*, E. *plantagineum*, and E. *expansum* extend to this Colony, the last three being found generally in moist or marshy places. Apium *prostratum* and A. *leptophyllum* are small herbs of the parsley kind, and Sium *latifolium* and S. *angustifolium* (whether introduced or not) are identical with the European species. Crantzia *lineata* is a small creeping herb extending to New Zealand and South America, and the species of Aciphylla (A. *simplicifolia* and A. *glacialis*) are limited to the mountainous regions of the south, being found at an elevation of from 5,000 to 7,000 feet above the sea-level. Daucus *brachiatus* is an herbaceous plant covered with short stiff hairs. The fruit is also bristly. Oreomyrrhis *andicola* is a species common to New Zealand and South America; and O. *pulvinifica* (a singular plant, which is described in the "Fragmenta" vol. viii, p. 185) is found on the Australian Alps at a great elevation. Sesili *Harveyanum* and S. *algens* are common to the high mountains of Victoria and New South Wales. There is some difficulty in determining the number of Umbellifers which have become naturalised. In addition to several species regarded as indigenous by Baron Mueller (such for instance as the two species of Sium), Mr. Bentham mentions Petroselinum *sativum* (Hoffm.), Ammi *majus* (Linn.), and Pastinaca *sativa* (Linn) ; whilst Anethum *fœniculum* (Willd.) and Bupleurum *rotundifolium* (Willd.) have spread on the banks of the Hawkesbury and in the neighbourhood of Parramatta. It appears from a review of Baron Mueller's second division of dicotyledonous plants that, exclusive of species known to have become introduced, there are in New South Wales 129 genera and 609 species.

I. DICOTYLEDONEÆ.

(III). SYNPETALEÆ PERIGYNEÆ.

The third division of dicotyledonous plants consists of such species as have, for the most part, connected petals, and these, as well as the stamens, are inserted on the tube of the calyx, and at a distance from the base of the ovary, or the stamens are affixed to the petals, whilst, with some few exceptions, the fruit is laterally adnate to the calyx. In this, as well as in the first division, several orders of monochlamydeous plants are incorporated, such

as the SANTALACEÆ, THYMELEÆ, and PROTEACEÆ, in the last of
which Baron Mueller regards " the floral envelope as homologous
to that of the closely allied LORANTHACEÆ, with an absence of a
calyx, comparable to the suppression of that organ in Diplolæna."

1. OLACINEÆ.—Of this order, which is widely dispersed over
the tropical and sub-tropical regions of the globe, none of the
genera are endemic in Australia, and the species extending to
New South Wales are only five. Olax *stricta* and O. *retusa* are
small shrubs with inconspicuous flowers and oblong drupes ;
Pennantia *Cunninghami* and P. *Endlicheri* (the latter from
Norfolk and Lord Howe's Islands) are larger shrubs, with corymbose flowers, and drupes half an inch long ; and Villaresia *Moorei*
is a lofty handsome tree, similar to the Javanese V. *suaveolens*.
Though in the Flora Australiensis this order stands amongst the
Thalamifloræ, Mr. Bentham remarks that it is very nearly allied
to the *Santalaceæ*, thereby confirming the views enunciated in the
Baron's Census.

2. Of the SANTALACEÆ four genera are limited to Australia,
and three are common to other parts. Thesium *Australe* is a
small plant with wiry stems, minute flowers, and little nuts
usually ribbed outside. Of Santalum or Sandalwood, the follow-
ing species occur for the most part on the western side of the
Dividing Range (S. *obtusifolium* only being found near the coast),
S. *lanceolatum*, S. *acuminatum*, S. *persicarium*, and S. *crassifolium*.
The second of these is the same as Fusanus *acuminatus*, or the
" Quandong," an edible fruit common to four of the Australian
Colonies. The genus Choretrum, represented by C. *glomeratum*,
C. *chrysanthum*, C. *spicatum*, C. *lateriflorum*, and C. *Candollei*,
consists of shrubs with slender and almost leafless branches and
minute flowers, and growing principally on ranges and mountains.
Leptomeria has L. *acida*, L. *Billardieri*, and L. *aphylla*, the first
of which is the " Native Currant" of the colonists, and can be
utilised for jam and acidulated drinks. Omphacomeria *acerba*
has the habit of the preceding genus, but differs in its unisexual
flowers and anthers with two distinct parallel cells. Exocarpus,
or the " Native Cherry" (so called because the nut appears to be
seated on the outside of the pericarps, but in reality on the
enlarged succulent pedicel), has the following species, E. *cupressi-
formis* (a graceful cypress-looking tree), E. *latifolia* (a tree with
oval oblong leaves), and the shrubs E. *odorata*, E. *spartea*, E.
aphylla (erect with stout, rigid, and finely-furrowed branches), E.
homoclada, from Lord Howe's Island, E. *phyllanthoides*, from
Norfolk Island, and E. *stricta*, an almost leafless plant with fruit
nearly globular.

3. LORANTHACEÆ, or the Mistletoe family, may be regarded
as parasitical (though in Australia there are two exceptions), with
thick fleshy leaves when present, and fruit a berry or drupe with

THE PLANTS OF NEW SOUTH WALES. 63

a single seed. The genus Viscum, which is familiar in England from V. *album,* the common mistletoe, is known in New South Wales by the species V. *angulatum* and V. *articulatum*—the former a leafless plant with opposite or forked branches; and the latter with flattened branches, articulate, and sometimes forked at almost every node. The flowers are very small, sessile, and clustered at the nodes. Notothixos *incanus* (including N. *subaureus* and N. *cornifolius*) occurs on the branches of Eucalypts, apples, and other trees, and parasitical on species of Loranthus—a parasite on a parasite! There is some difficulty about the species of Loranthus, because it is supposed that the same species may be subject to a considerable amount of variation according to the nature of the tree on which it grows. The following are reckoned for different parts of the Colony :—L. *celastroides,* L. *myrtifolius,* L. *longiflorus,* L. *dictyophlebus,* L. *alyxifolius,* L. *exocarpi,* L. *maytenifolius,* L. *linophyllus,* L. *pendulus,* and L. *quandang.* Some of these parasites have rather showy flowers, and as they hang from the bush trees they have an interesting appearance amidst the diversity of foliage with which they are associated. Atkinsonia *ligustrina* (a plant ever to be connected with the memory of the late Mrs. Calvert, better known as Miss Louisa Atkinson) was originally regarded as a species of Nuytsia, but it has been separated from that genus by Baron Mueller, because it differs in the shape of its fruit and the character of its inflorescence, and, therefore, it is now regarded as a living memorial of a lady who, by her various collections and artistic skill, contributed materially to the natural history of the Blue Mountains. A. *ligustrina* is not parasitical on other trees, but it is seen near Mount Tomah as a terrestrial plant, attaining a height of 4 or 5 feet, with opposite lanceolate leaves, axillary racemes of flowers, and scarlet fruits. A full description of this interesting shrub is given in the "Fragmenta Phytographiæ Australiæ," vol. v, p. 34. The true Nuytsia (N. *floribunda*) is, like the preceding, limited to a single species, endemic in Australia, and extends from King George's Sound to the Swan and Murchison Rivers. This tree attains a height of 30 or 40 feet, with orange-coloured flowers in showy racemes, and dry drupes with three broad longitudinal wings, thus differing from Atkinsonia, which has wingless fruits. It should be mentioned in connection with the Loranthaceæ that several of the species afford fruits agreeable to the aboriginal natives. One of our Loranths seems to grow from the trunks of trees, and to extend 30 or 40 feet up the branches.

4. PROTEACEÆ.—This large and interesting order, which for all Australia numbers nearly 600 species, is a very characteristic one in New South Wales; for although with two exceptions the genera are found to extend on the one hand to New Caledonia, the Indian Archipelago, and tropical Asia, and on the other to

South America, yet they have one of the principal seats in Australia, and form in some parts of the continent a peculiar feature of the vegetation. The eminent botanist R. Brown, as writers have frequently observed, was very happy in giving the name to the order, for, whilst the species agree in many well-defined marks, they are found to differ most widely in their appearance, so that the casual observer would scarcely imagine that they belonged to the same family. And yet when the woody texture of the leaves, the irregular tubular calyxes, and the stamens (usually four), which are placed on the lobes of the calyx, are taken into consideration, it is almost impossible to mistake the species for those of any other order. At a time when the Australian Proteaceæ were but little known in Europe the order was regarded more for the beauty and singularity of its flowers than for any particular usefulness. Now, however, that the species have become better known, and their properties duly investigated, it is found that the order comprises trees which can be utilised for the value of their wood and the flavour of their fruit ; whilst some of the flowers yield saccharine juices which are attractive to bees and small birds. The genera may be divided into two sub-orders :—(1) Numentaceæ, or such as have indehiscent nuts or drupes ; and (2) Folliculares, or such as have dehiscent fruit, follicular or two-valved.

1. NUMENTACEÆ.

The genus Petrophila, which is very abundant in West Australia, is limited to three species in New South Wales—P. *pedunculata*, P. *pulchella*, and P. *sessilis*. The flowers of these shrubs are white or yellow, in dense spikes, and the leaves rigid twice or thrice pinnate.

Isopogon is similar in habit to Petrophila, with twenty-five species in W Australia, and four (I. *anethifolius*, I. *petiolaris*, I. *anemonifolius*, and I. *ceratophyllus*), near Sydney and beyond the Blue Mountains.

Conospermum is limited to Australia, and the greater number of species are in West Australia. The species in East Australia are small shrubs with white or bluish flowers and entire leaves, C. *longifolium*, C. *tenuifolium*, C. *patens*, C. *taxifolium*, C. *ericifolium*, C. *ellipticum*, and C. *stæchadis*.

Symphyonema has only two species, and these peculiar to East Australia, S. *montanum* and S. *paludosum*, small shrubs with yellow flowers in slender spikes and leaves trichotomously divided.

Persoonia is a large genus, the species of which are nearly equally divided between East and West Australia, those of the former being P. *ferruginea*, P. *media*, P. *cornifolia*, P. *marginata*, P. *sericea*, P. *Mitchellii*, P. *fastigiata*, P. *hirsuta*, P. *chamœpithys*,

P. salicina, P. prostrata, P. lanceolata, P. lucida, P. linearis, P. pinifolia, P. Caleyi, P. ledifolia, P. revoluta, P. mollis, P. rigida, P. curvifolia, P. oblongata, P. Cunninghami, P. myrtilloides, P. oxycoccoides, P. nutans, P. angulata, P. virgata, P. chamæpeuce, P. juniperina, P. tenuifolia, and *P. acerosa.* These sixteen species differ in size and in the shape of their leaves, but the flowers, for the most part yellow, are very similar, and the fruits, which are succulent, are generally known by the native name of "geebungs."

2. FOLLICULARES.

Macadamia *ternifolia,* "the Queensland Nut," is a tree of moderate size having finely-grained wood capable of taking a good polish, and yielding an edible fruit of excellent flavour.

Helicia *Youngiana,* H. *præalta,* H. *glabrifolia,* and H. *ferruginea* are closely allied to the preceding, but differing in the foliage and fruit.

Xylomelum *pyriforme* is the well-known "wooden pear" of the colonists. It does not attain any size, but the wood is prettily marked and available for rough furniture.

Lambertia *formosa,* sometimes called "the honey flower," is the only species of a genus numbering eight species in Western Australia. It is a tall shrub with red flowers and linear pungent leaves.

Orites *excelsa* is one of the largest and handsomest of proteaceous trees, common to many of the rivers of New South Wales, with varying leaves, and flowers in axillary spikes.

Strangea *linearis* (Grevillea *strangea,* Benth.) is a small subtropical shrub, differing from Grevillea principally in the accessory wing covering the seed (see Fragmenta, vol. vii, p. 131).

Grevillea is a large genus containing nearly 160 species for all Australia, the most of which are peculiar to West Australia, thirty-nine only occurring in East Australia. Of these, some are mere shrubs, and others are handsome trees yielding fine timber. G. *robusta,* "the silky oak," is valued not only as an ornamental tree, but for its wood, which is extensively used for staves to tallow-casks. G. *Hilliana* is also a beautiful tree, with wood easily worked. In addition to these the following species may be enumerated, several of which are seen in cultivation and much admired :—G. *pterosperma,* G. *Caleyi,* G. *asplenifolia,* G. *laurifolia,* G. *Gaudichaudi,* G. *acanthifolia,* G. *floribunda,* G. *cinerea,* G. *montana,* G. *obtusiflora,* G. *arenaria,* G. *mucronulata,* G. *Baueri,* G. *lanigera,* G. *rosmarinifolia,* G. *juncifolia,* G. *lavandulacea,* G. *Huegelii,* G. *striata,* G. *buxifolia,* G. *phylicoides,* G. *sphacelata,* G. *Victoriæ,* G. *punicea,* G. *oleoides,* G. *trinervis,* G. *juniperina,* G. *sericea,* G. *capitella,* G. *leiophylla,* G. *linearis,* G. *parviflora,* G. *australis,* G. *triternata,* G. *ramosissima,* G. *nematophylla,* G. *anethifolia,* and G. *Miqueliana.* Hakea is a genus

E

closely allied to Grevillea, and limited to Australia. There is
scarcely any character to distinguish the two genera excepting
the seed-wing, and that cannot always be relied on. The species
for New South Wales are—H. *Fraseri*, H. *eriantha*, H. *pugioniformis*, H. *Pampliniana*, H. *saligna*, H. *purpurea*, H. *gibbosa*, H.
propinqua, H. *acicularis*, H. *cycloptera*, H. *microcarpa*, H. *dacty-
loides*, H. *ulicina*, and H. *flexilis*. Some of the Western Australian species are worthy of cultivation, especially for shrubberies.

Stenocarpus *sinuatus*, or " tulip-tree," is a handsome tree of
moderate size (40 or 50 feet), with umbels of bright red flowers
and pinnatifid leaves a foot long. The wood is nicely marked.
S. *salignus*, or " beef-wood," is a fine tree, with wood of a red
colour, and valued for cooper's work.

Lomatia is a genus represented in the mountains of extra-
tropical South America, as well as in Australia. The leaves vary
considerably in the species, but there is great uniformity in the
structure of the flowers. L. *longifolia* sometimes attains a
height of 20 or 30 feet, but L. *ilicifolia*, L. *Fraseri*, and L. *silai-
folia* are mere shrubs.

Telopea *speciosissima*, " the native tulip or waratah," with its
large crimson flowers in dense ovoid or globular heads, is regarded
by many as the handsomest of Australian flowers. T. *oreades*
is common to Victoria and New South Wales, and Tasmania has
T. *truncata*. The genus does not extend to the other colonies.

Banksia (so named in honor of Sir Joseph Banks, the friend
of Cook, and the patron of Australian botanists) has only eight
species in New South Wales ; the remaining species (thirty-six)
belong to West Australia. The eastern ones are B. *ericifolia*,
B. *spinulosa*, B. *collina*, B. *marginata*, B. *integrifolia*, B. *latifolia*,
B. *serrata*, and B. *æmula*. The larger of these, under the popular
name of " honeysuckle," are used for the knees of boats, &c.
The genus Banksia is endemic in Australia, but the distribution
is remarkable, two only occurring in S. Australia, two in Tasmania, five in Victoria, five in Queensland, and one in N. Australia.
The allied genus Dryandra, differing from Banksia chiefly in the
involucre surrounding the flowers, is endemic in and limited to
W. Australia, none of the forty-seven species extending to the
other colonies.*

5. THYMELEÆ.—This family of plants is common at the Cape
of Good Hope and in Australia. The whole number of species
in the latter country is between seventy and eighty, of which
twenty-seven extend to New South Wales. In the genus Pimelea
alone twenty-five of these may be reckoned, some of which are

---

* Baron Mueller has added another genus to the PROTEACEÆ in *Hicks-
beachia* (H. *pinnatifolia*), which is allied to *Macadamia*, and occurs on the
Tweed.

regarded as ornamental shrubs in cultivation, rising to the height of 3 or 4 feet, and producing numerous heads of flowers of a white, pink, or red colour. The species near Sydney (P. *linifolia*, P. *curviflora*, and P. *spicata*) are not very conspicuous ; but amongst the following, from beyond the Dividing Range and the northern parts of the Colony, some are interesting, viz. :—P. *alpina*, P. *longifolia*, P. *glauca*, P. *colorans*, P. *collina*, P. *spathulata*, P. *ligustrina*, P. *humilis*, P. *simplex*, P. *sericostachya*, P. *trichostachya*, P. *axiflora*, P. *microcephala*, P. *pauciflora*, P. *serpyllifolia*, P. *flava*, P. *petrophila*, P. *hirsuta*, P. *altior*, P. *octophylla*, P. *petræa*, P. *phylicoides*, P. *stricta*, and P. *penicillaris*. Two of the most admired in cultivation (P. *spectabilis* and P. *suaveolens*) are western species. Drapetes *Tasmanica* is a prostrate, intricately tufted, little plant, with flowers in small terminal heads and a minute ovate fruit. This is an alpine species growing on Mount Kosciusko and elsewhere, at a considerable elevation above the sea. Wickstræmia *Indica* is found in the immediate vicinity of Port Jackson. It is a low shrub with octandrous flowers of a greenish colour, and red berry-like fruits.

6. CORNACEÆ.—This order is represented in Australia by a solitary species common to the islands of the South Pacific. In the " Flora Australiensis " this tree is called Marlea *vitiensis*; but according to the Census of the Baron, who desires to restore the original name given to it in 1790, the species is designated Stylidium *vitiense*. The flowers are in short axillary cymes, but the leaves are sometimes 5 inches long.

7. RUBIACEÆ is an order well characterised by epipetalous stamens, straight anthers bursting lengthways, and opposite leaves, with interpetiolar stipules. Some of the species are worthy of cultivation for the beauty of their flowers, whilst others possess medicinal and industrial properties. Of the genus Oldenlandia (Hedyotis Benth.), only one species belongs to New South Wales, and that is found beyond the Darling (O. *tillæacea*). Dentilla *repens* is a small prostrate herb found on the Blue Mountains. Gardenia is limited to North Australia and Queensland, but the allied genus Randia (to which G. *chartacea* is now referred) has the following species in the Northern districts :— R. *chartacea*, R. *Benthamia*, R. *Moorei*, R. *stipularis*, and R. *densiflora*. Ixora and Hodgkinsonia have each one species, I. *Becklerii* and H. *ovatiflora*; but Canthium (a genus more widely distributed in New South Wales) has C. *latifolium*, C. *lucidum*, C. *oleifolium*, C. *buxifolium*, C. *vaccinifolium*, and C. *coprosmoides*, some of which occur on the Blue Mountains, and others beyond the Dividing Range, or on the Northern rivers. Morinda *jasminoides* is a straggling plant found generally in creeks, and Coelospermum *paniculatum* is a woody climber on the Clarence River and elsewhere to the north, with white flowers

in dense panicles. Psychotria has P. *loniceroides* near Sydney, and P. *daphnoides* and P. *Fitzalani* farther north. These are all endemic. Coprosma *Billardieri*, C. *hirtella*, C. *Baueri*, C. *putida*, C. *pilosa*, C. *lanceolata* are found for the most part in mountainous ranges, and, though one of them has edible berries, the plants are generally characterised by an unpleasant scent. The same may be said of Opercularia, of which O. *scabrida*, O. *aspera*, O. *hispida*, O. *diphylla*, O. *ovata*, and O. *varia* are herbs or small shrubs common on this side of the Blue Mountains. This is a genus peculiar to Australia, and the species are somewhat variable. Pomax *umbellata* is a singular little plant, differing from Opercularia only in the umbelliferous nature of its inflorescence. Spermacoce is found principally in Queensland and Northern Australia, but S. *multicaulis* passes the borders of the former into New South Wales. With this order two species of Asperula (A. *geminifolia* and A. *oligantha*) and two of Galium (G. *umbrosum* and G. *Australe*) are now associated. These are small herbs with quadrangular stems and leaves in whorls of four to eight. Sherardia *arvensis* (Willd.) and Galium *aparine* (Linn.) are introduced weeds.

8. The order of CAPRIFOLIACEÆ is a small one in New South Wales, Sambucus *xanthocarpa* and S. *Gaudichaudiana*, two species of elder, being the only plants of the genus known in the Colony.

9. The beautiful order of PASSIFLORÆ, or Passion-flowers, is confined to one genus in Australia, and the four species (P. *Herbertiana*, P. *cinnabarina*, P. *glabra*, and P. *aurantia*) of Eastern Australia cannot be compared with those of America, either as ornamental plants or as producing edible fruit. They are climbers with axillary tendrils, succulent fruit, and lobed leaves; the flowers are greenish, white, red, or orange. P. *cœrulea* (Willd.), an introduced species, is spreading very much on old buildings, fences, &c.

10. The Gourd or Melon family (CUCURBITACEÆ) has not many species in New South Wales, and those are not, so far as known, of much practical utility. Trichosanthes *palmata* is a coarse climber, common to India, Queensland, and New South Wales, but it does not extend farther south than the Tweed River. Cucumis *trigonus* and C. *chate* (two native cucumbers) have small globular or ovoid fruits, and one of them is eaten in great quantities by the aboriginal natives. Momordica *balsamina* is a slender climber with yellow flowers and red fruit. This is a plant widely spread in Asia and Africa, and seen occasionally in cultivation. Bryonopsis *laciniosa* and B. *Pancheri* are plants of a similar character, with small flowers and yellow globular fruit. Melothria *Cunninghamii*, M. *Baueriana*, M. *Muelleri*, and M. *Maderaspatana*, kind of native melons, occur beyond the Dividing

Range or in the warmer parts of the Colony. Sicyos *angulata* (the only species of the order growing near Sydney) is a slender climbing plant, with small whitish flowers, and fruit covered with barbed prickles.

11. The Composite family of plants (COMPOSITÆ of D'Candolle and ASTERACEÆ of Lindley) is one of the most natural and widely distributed families in the vegetable kingdom. It is distinguished from all others by an inferior ovary with a single ovule, filiform style usually divided at the top, syngenesious stamens and capitate flowers. The species, of which about 10,000 are known, are found in every part of the world, and of these between 500 or 600 occur in Australia. All plants of what are called the Daisy and everlasting kind belong to this order, and although many of them are insignificant herbs or mere weeds, yet in Australia a few are large shrubs or trees. There is great difficulty in the grouping of the genera ; but Mr. Bentham remarks that the minute differences in the shape of the style branches are the most reliable for that purpose. The following is the arrangement in Baron F. von Mueller's Census :—

(1.) Centratherum : C. *muticum*, a rigid spreading herb, with purple florets, common to tropical America, the Philippine Islands, and Australia.

(2.) Vernonia : V. *cinerea*, a weed of tropical Asia, with purplish florets, found in New South Wales, Queensland, and N. Australia.

(3.) Adenostemma : A. *viscosum*, a herb with small flower-heads, usually glandular-pubescent. This is a weed frequent in the warmer parts of the globe.

(4.) Eupatorium : E. *cannabinum* is very common in the temperate regions of the Northern Hemisphere, and is probably an introduced plant.

(5.) Lagenophora: L. *Billardieri*, L. *solenogyne*, and L. *emphysopus* are small herbs of the Daisy kind, remarkable for their compressed seeds, which are contracted at the top with a distinct neck.

(6.) Brachycome, or the Australian Daisy, is represented in New South Wales by the following species :—B. *diversifolia*, B. *segmentosa*, B. *melanocarpa*, B. *radicans*, B. *goniocarpa*, B. *pachyptera*, B. *microcarpa*, B. *Stuartii*, B. *scapigera*, B. *graminea*, B. *angustifolia*, B. *linearifolia*, B. *basaltica*, B. *trachycarpa*, B. *exilis*, B. *ptychocarpa*, B. *debilis*, B. *decipiens*, B. *cardiocarpa*, B. *nivalis*, B. *scapiformis*, B. *stricta*, B. *heterodonta*, B. *ciliaris*, B. *calocarpa*, B. *chrysoglossa*, B. *marginata*, B. *Sieberi*, B. *discolor*, B. *multifida*, and B. *collina*, in all thirty-one species. This genus is nearly allied to Bellis, or the true daisy. The flowers are terminal, and the ray is white, blue, or purple. Some of the species are worthy of cultivation.

(7.) Minuaria : M. *leptophylla*, M. *Cunninghami*, M. *Candollei*, and M. *suædifolia* are small shrubs indigenous on the west side of the Dividing Range. The genus is limited to Australia.

(8.) Calotis : This is another genus of the daisy kind, differing, however, in the barbed bristle of the pappus, and occasionally in the yellow colouring of the florets. The species are C. *dentex*, C. *cuneifolia*, C. *glandulosa*, C. *cymbacantha*, C. *erinacea*, C. *scab-iosifolia*, C. *scapigera*, C. *lappulacea*, C. *microcephala*, C. *plumifera*, C. *porphyroglossa*, and C. *hispidula*.

(9.) Aster (Olearia of Bentham) : This is a large genus, and comprises not only herbaceous plants but large shrubs, or even trees, of which the musk-tree (A. *argophyllus*) rises sometimes to the height of 25 feet and upwards. The flowers of the genus vary considerably in size, and the leaves of some species have white and stellate hairs on their under surface. The species occurring in New South Wales are numerous, but few of them are found near the coast. They are—A. *megalophyllus*, A. *chryso-phyllus*, A. *alpicola*, A. *rosmarinifolius*, A. *oliganthemus*, A. *argo-phyllus*, A. *cydonifolius*, A. *myrsinoides*, A. *dentatus*, A. *stellulatus*, A. *asterotrichus*, A. *gravis*, A. *Nerustii*, A. *Siemssii*, A. *tubuli-florus*, A. *axillaris*, A. *ramulosus*, A. *florulentus*, A. *microphyllus*, A. *Mitchelli*, A. *cyanodiscus*, A. *pimeleoides*, A. *iodochrous*, A. *conocephalus*, A. *calcareus*, A. *magniflorens*, A. *Muelleri*, A. *decur-rens*, A. *glutescens*, A. *orarius*, A. *teretifolius*, A. *ellipticus*, A. *glandulosus*, A. *Cunninghami*, A. *adenophorus*, A. *exul*, A. *Huegelii*, A. *Balli*, A. *Moorei*, and A. *celmisia*.

(10.) Vittadinia : V. *australis* and V. *scabra* are herbs with a yellow centre, blue ray, and hirsute leaves. The former is more frequent near Sydney, and is very variable in its foliage.

(11.) Podocoma : P. *cuneifolia*, a herb with large flower-heads, extending from South Australia into New South Wales.

(12.) Erigeron : E. *pappochromus* and E. *conyzoides* are indige-nous, but E. *linifolius* and E. *Canadensis* ("Cobbler's Pegs") are weeds of foreign origin.

(13.) Conyza : C. *viscidula* is allied to the preceding genus, and may have come originally from India.

(14.) Pluchea : P. *Eyrea* is a shrub with small corymbose flowers and decurrent leaves, found on the Darling.

(15.) Epaltes : E. *Cunninghami*, E. *Australis*, and E. *pleiocheta* are insignificant herbs, growing generally in marshy places.

(16.) Pterocaulon (Mononteles, Labill.) : P. *Billardieri* and P. *sphacelatus* are under-shrubs, usually glandular and strongly scented, with leaves decurrent on the stem. The first of these has been collected in New Guinea. (F. v. M., Papuan Plants.)

(17.) Stuartina : S. *Muelleri* is the only species of the genus, and limited to Australia. It is a small diffuse annual, only a few inches high, and extending from Victoria to the southern part of New South Wales.

(18.) Gnaphalium: The species of this genus are weeds with small flower-heads, and more or less cottony or woolly. Some of the following may have been introduced:—G. *luteo-album*, G. *Japonicum*, G. *purpureum*, G. *Indicum*, G. *indutum*, and G. *Traversii*.

(19.) Antennaria: A. *uniceps* is a small densely-tufted herb, limited to the elevated mountains of Victoria and the southern ones of New South Wales.

(20.) Leontopodium: L. *catipes* (Raoulia, Flo. Aust.) is another alpine species, forming large tufts of a silvery-white appearance.

(21.) Podotheca: P. *angustifolia* is a coarse herb with stipitate seeds, occurring on both sides of the Murray.

(22.) Ixioloena: I. *brevicompta*, I. *leptolepis*, and I. *tomentosa* are herbs of little beauty in the W. and N.W. interior.

(23.) Podolepis: P. *rhytidochlamys*, P. *longipedata*, P. *acuminata*, P. *canescens*, P. *Lessoni*, and P. *Siemssenia* are herbs with terminal flower-heads, varying in colour. P. *acuminata*, with large flower-heads, is the only species near Sydney.

(24.) Anthrixia: A. *tenella* is a small branching annual, only an inch or two in height, with white florets.

(25.) Leptorrhynchus : L. *squamatus*, L. *panoetioides*, L. *pulchellus*, L. *elongatus*, L. *medius*, L. *Waitzia*, and L. *nitidulus* may be termed "everlasting flowers," seldom exceeding a foot in height, having yellow flowers, and involucral bracts sometimes descending along the stem.

(26.) Waitzia: W. *corymbosa* is a scabrous plant with numerous flower-heads in a dense corymb, varying in colour from pale to a dark yellow, white, or bright pink. This is the only species of the genus which extends to East Australia, the others being limited to West Australia.

(27.) Helipterum, with the genus which follows it, contains many species called "Everlastings" or "Immortelles," because, from the dryness of their nature, they last so much longer than flowers in general. The two genera, Helichrysum and Helipterum, differ very little from each other except in the plumose pappus of the latter. Some of the species, such as Helichrysum *bracteatum* and H. *elatum*, are admired in cultivation, the former being a handsome plant and varying in colour. The western Helipterum *Manglesii* is also a very elegant plant. The eastern species are: —H. *anthemoides*, H. *polygalifolium*, H. *floribundum*, H. *incanum*, H. *cotula*, H. *hyalospermum*, H. *polyphyllum*, H. *strictum*, H. *corymbiflorum*, H. *pygmoeum*, H. *moschatum*, H. *Tietkensii*, H. *baccharoides*, H. *pterochoetum*, H. *exiguum*, and H. *dimorpholepis*.

(28.) Helichrysum : The species of New South Wales are H. *semifertile*, H. *Baxteri*, H. *rutidolepis*, H. *scorpioides*, H. *obtusifolium*, H. *Calvertianum*, H. *lucidum*, H. *elatum*, H. *leucopsidium*, H. *Blandowskianum*, H. *oxylepis*, H. *collinum*, H. *podolepidium*,

H. *ambiguum*, H. *apiculatum*, H. *semipapposum*, H. *Dockerii*, H. *obovatum*, H. *Bidwillii*, H. *Becklerii*, H. *diotophyllum*, H. *diosmifolium*, H. *retusum*, H. *decurrens*, II. *Cunninghami*, H. *cinereum*, H. *rosmarinifolium*, H. *ferrugineum*, H. *obcordatum*, H. *pholidotum*, and H. *cuneifolium*.

(29.) Cassinia: The species are designated by European gardeners "neat New Holland shrubs, with white or yellow flowers." The flower-heads are rather small, numerous, and arranged in corymbs or panicles.—C. *leptocephala*, C. *compacta*, C. *denticulata*, C. *longifolia*, C. *aurea*, C. *aculeata*, C. *tenuifolia*, C. *lœvis*, C. *quinquefaria*, C. *arcuata*, C. *subtropica*, and C. *Theodori*.

(30.) Humea: H. *elegans* is a strongly-scented plant, rising to the height of several feet, with a large loose panicle of small flowers, and long stem-clasping leaves. H. *ozothamnoides* is a smaller plant growing on the banks of the Murray.

(31.) Acomis: A. *rutidosis* is a slender herb limited to the northern rivers of the Colony.

(32.) Rutidosis: The following species differ from Helichrysum and Podolepis principally in the chaffy scales of the pappus :—R. *leptorrhynchoides*, R. *helichrysoides*, and R. *pumilo*.

(33.) Ammobium: A. *alatum* and A. *craspedioides* are chiefly remarkable for their winged stems or decurrent leaves; the former is a cultivated plant, and regarded in Europe as half-hardy.

(34.) Millotia: M. *tenuifolia* and M. *Greevesii* are insignificant annuals, with small flower-heads, found in arid parts of the interior.

(35.) Toxanthus *perpusillus*: A dwarf annual on the banks of the Murray.

(36.) Eriochlamys: E. *Behrii*, a very small woolly plant, only a few inches high, and extending from the Lachlan and Darling to the Barrier Range.

(37.) Myriocephalus: M. *rhizocephalus* is a small tufted annual, with clusters of flower-heads, enclosed in a general involucre, the bracts being transparent, with green mid-ribs, ciliate with long woolly hairs. M. *Stuartii* is a larger plant, the heads of which are not so woolly.

(33.) Angianthus: A. *tomentosus*, A. *brachypappus*, A. *pusillus*, A. *Pressianus*, and A. *strictus* are small herbaceous plants, with flower-heads on a cylindrical receptacle. They are limited to the dry regions of the interior.

(39.) Gnephosis: G. *eriocarpa*, G. *skirrophora*, G. *arachnoides*, and G. *cyathopappus* are very similar in character and habit to the preceding.

(40.) Calocephalus: This genus differs but little from Gnephosis. The clusters of the flower-heads are white or yellow. C.

Brownii, C. *Sonderi*, C. *lacteus*, C. *citreus*, and C. *platycephalus* are chiefly from the Lachlan and Darling, one species only reaching to New England and Argyle.

(41.) Gnaphalodes *uliginosum* is a very small plant, with flower-heads surrounded by woolly leaves. It is found on the Lachlan.

(42.) Craspedia: C. *Richea*, C. *pleiocephala*, C. *chrysantha*, and C. *globosa* are plants with large globular or ovoid clusters of flower-heads of a yellow colour, and with few cauline leaves. The first is common near Sydney.

(43.) Chthonocephalus *pseudoevax* is a minute annual with scarcely any stem, and numerous heads of sessile flowers.

(44.) Siegesbeckia *orientalis* is a weed widely dispersed over the warmer regions of the world.

(45.) Enhydra *paludosa* is a species similar to that in India. It grows in the water, has opposite leaves, and flower-heads sessile in the forks of the stem.

(46.) Eclipta *alba* and E. *platyglossa* are weeds found principally in marshy places, the one with white and the other with yellow flowers, with opposite leaves, and sprinkled over with rough hairs.

(47.) Wedelia: W. *spilanthoides* and W. *biflora* have larger flowers than those of Eclipta. W. *hispida*, a coarse plant with large yellow flowers, has spread very much of late on the banks of the Hawkesbury.

(48.) Spilanthes *grandiflora* is a plant with flowers on long stalks, having the ray florets very long. It occurs on the Clarence and Richmond, and also in New England.

(49.) Bidens *tripartitus*, B. *pilosus*, and B. *bipinnatus* have probably been introduced.

(50.) Glossogyne *tenuifolia* has small flower-heads and leaves pinnately divided. It is found in North and East Australia, as well as in New Caledonia and the Indian Archipelago.

(51.) Soliva *anthemifolia* (a diffuse herb, with flower-heads closely sessile) is probably of foreign origin.

(52.) Cotula: This genus has five species in New South Wales—C. *filifolia*, C. *coronopifolia*, C. *Australis*, C. *alpina*, and C. *filicula*. The second is a common weed in swampy places, and the third may be found almost anywhere in cultivated ground or about sheep-stations.

(53.) Centipeda: C. *racemosa*, C. *orbicularis*, C. *Cunninghami*, and C. *thespesioides* are herbs with toothed leaves. Some of these have sternutatory properties, and snuff has been prepared from them under the direction of Baron Mueller. The third species is figured by him amongst his lithograms of Victorian Plants.

(54.) Elachanthus *pusillus* is a small weed on the Darling.

(55.) Isoetopsis *graminifolia* is another little herb on the Lachlan and Darling, having a number of small flower-heads crowded within the radical leaves.

(56.) Gynura *pseudochina* is a succulent herb, with corymbs of yellow flowers, found only in the northern parts of the Colony.

(57.) Senecio is the largest genus of Composites, and the species of it occur in nearly every part of the globe. Mr. Bentham remarks, that of the twenty-eight Australian species one only extends to New Zealand (S. *lautus*), the rest being endemic. One of the largest species, S. *Bedfordii* (Bedfordia *salicina*, D.C.) is common to Tasmania, Victoria, and the southern part of New South Wales. The species are S. *Gregorii*, S. *platylepis*, S. *pectinatus*, S. *spathulatus*, S. *magnificus*, S. *insularis*, S. *macranthus*, S. *Daltoni*, S. *lautus*, S. *vagus*, S. *amygdalifolius*, S. *velleioides*, S. *australis*, S. *Behrianus*, S. *brachyglossus*, S. *Georgianus*, S. *odoratus*, S. *Cunninghami*, S. *anethifolius*, and S. *Bedfordii*.

(58.) Erechtites is nearly allied to the preceding, and differs only in having filiform female florets. The species of New South Wales are E. *prenanthoides*, E. *Atkinsoniæ*, E. *arguta*, E. *mixta*, E. *quadridentata*, and E. *hispidula*. These, as well as most of the species of Senecio, bear some resemblance to the European groundsel (S. *vulgaris*).

(59.) Cymbonotus *Lawsonianus* is a stemless herb with yellow flowers, and leaves cottony white on the under surface.

(60.) Saussurea *carthamnoides*, common to India, China, Japan, and the East Coast of Australia, is a rigid annual with pinnatifid leaves and large flower-heads of purple florets.

(61.) Centaurea *Australis* (Leuzea D.C.) is an erect herb with large solitary purple flowers on a long peduncle.

(62.) Microseris *Forsteri* is a perennial with fleshy roots, yellow florets, and pinnatifid leaves.

(63.) Crepis *Japonica* is a small annual with flower-heads in loose corymbs or panicles, probably introduced from India or China.

(64.) Blumea *hieracifolia*, a very common species in Asia, extends to New South Wales.

From this review of the Composites, it appears that in New South Wales sixty-four genera are represented by 279 species. In addition to these, the introduced species are very numerous, consisting for the most part of weeds brought from different parts of the world, and scattered plentifully throughout the country. There may be some difference of opinion as to whether certain species are indigenous or not, but the following are admitted to be of foreign origin :—

Centaurea *melitensis* (Linn.), C. *calcitrapa* (Linn.), Onopordon *acanthium* (Linn.), Cirsium *lanceolatum* (Scop.), Carduus *Marianus* (Linn.), Erigeron *Canadensis* (Linn.), E. *linifolius* (Willd.),

Xanthium *spinosum* (Linn.), Tagetes *glandulifera* (Schr.),
Anthemis *nobilis* (Willd.), A. *cotula* (Linn.), Aster *dumosus*
(Willd.), Chrysanthemum *segetum* (Linn.), C. *Parthenium*
(Pers.), Galinsoga *parviflora* (Cav.), Gnaphalium *luteo-album*
(Linn.), G. *purpureum* (Linn.), Cryptostemma *calendulaceum*
(R. Br.), Hypochœris *glabra* (Linn.), H. *radiata* (Linn.),
Sonchus *oleraceus* (Linn.), Cichorium *Intybus* (Linn.), Senecio
scandens (D.C.), S. *vulgaris* (Linn.), Leontodon *hirtus* (Linn.),
Tragopogon *porrifolius* (Linn.), Taraxacum *dens-leonis* (Desf.),
Tolpis *barbata* (Willd), Picris *hieracioides* (Linn.), Wedelia
hispida (Kth.)

12. CAMPANULACEÆ, or the Bellworts, are not so numerous in
this part of the world as in the Northern Hemisphere and in
South Africa, but amongst the species some are pretty plants and
worthy of cultivation. The flowers are mostly blue, equal or
unequal in the lobes of the corolla and pentandrous. Of Lobelia
the following extend to the eastern coast :—L. *Browniana*, L. *sim-
plicicaulis*, L. *microsperma*, L. *dentata*, L. *gracilis*, L. *trigonocaulis*,
L. *anceps*, L. *purpurascens*, L. *pratioides*, L. *gelida*, L. *concolor*,
L. *Benthami*, L. *pedunculata*, and L. *debilis*. The genus Isotoma
differs from Lobelia only in having the corolla nearly equally
lobed, the tube cylindrical and entire, and the stamens inserted
near the summit of it. I. *fluviatilis* is common in moist places on
the banks of creeks near Sydney ; but I. *axillaris* and I. *petræa*,
are beyond the range. The first of these varies in the shape of
the leaves and the colour of the flowers from white to lilac.
Wahlenbergia *gracilis*, the "Australian Blue Bell," is the only
one of the genus frequent in all the Australian Colonies, and
varying very much in size.

13. CANDOLLEACEÆ or STYLIDEÆ (R. Brown) : It appears that,
according to priority of nomenclature, the first name should be
adopted in honor of the eminent D'Candolle, as it was given by
Labillardiere in 1805, whilst Brown's name, which, however, is
very appropriate, was not bestowed on the order until 1810. The
order is a very singular one, the structure of the column into
which the stamens and style are blended being, according to
Lindley, different from anything in the vegetable kingdom, except
in orchids. The column is very irritable, and in dry weather
springs up when touched. Candollea or Stylidium is abundant
in W. Australia, having sixty-three species, and nearly all of
them limited to that colony. In New South Wales only the
following occur:—C. or S. *graminifolium*, C. or S. *lineare*, C. or
S. *debile*, C. or S. *laricifolium*, C. or S. *despectum*, and C. or S.
eglandulosum. Leewenhoekia *dubia* is a minute, glandular,
pubescent plant, 1 or 2 inches high, and found in swampy
places beyond the range. In this the irritability resides in the
labellum, whilst the column is immovable.

14. The order GOODENIACEÆ, which is nearly limited to Aus
tralia (a few species only inhabiting India, &c.), is characterised
by the peculiarity of the stigma, which is seated at the bottom o
a cup or covering called an indusium. This organ is intimately
connected with the impregnation of the flowers, and needs care
ful observation. The flowers are yellow, blue, or purple, pentan
drous, and for the most part irregular. Brunonia *australis* is
common to all the Australian Colonies. Dampiera is principally
a western genus, twenty-nine species being almost peculiar to
Western Australia. The following species are indigenous in this
Colony :—D. *Brownii*, D. *lanceolata*, D. *marifolia*, D. *rosmarinifolia*, D. *stricta*, D. *Scottiana*, and D. *adpressa*. Scævola (the
" left-handed or oblique flower") is also for the most part a
western genus, but these species, S. *spinescens*, S. *hispida*, S.
Hookeri, S. *suaveolens*, S. *ovalifolia*, S. *æmula*, and S. *microcarpa*,
are found in Eastern Australia. Selliera *radicans* is remarkable for its distribution, being found in New Zealand, South
America, and Australia. Goodenia has twenty-six species in
Western Australia, and the following in New South Wales :—
G. *decurrens*, G. *bellidifolia*, G. *stelligera*, G. *ovata*, G. *varia*, G.
barbata, G. *geniculata*, G. *hederacea*, G. *heterophylla*, G. *glabra*,
G. *rotundifolia*, G. *calcarata*, G. *grandiflora*, G. *cycloptera*, G.
elongata, G. *pinnatifida*, G. *heteronema*, G. *glauca*, G. *humilis*, G.
paniculata, and G. *gracilis*.

These plants, with the exception of G. *ovata*, which is sometimes a good-sized shrub, are herbaceous. The flowers are
generally yellow, except those of G. *barbata*. G. *heterophylla*
varies very much in foliage, changes colour in drying, and occasionally has its stem fasciated, or very much flattened, as in the
cockscomb. Velleya resembles Goodenia, but it differs in having
a free calyx and dichotomous inflorescence. The eastern species
are more numerous than the western ones, being—V. *connata*,
V. *perfoliata*, V. *paradoxa*, V. *lyrata*, V. *macrocalyx*, V. *spathulata*,
and V. *montana*.

From a consideration of Baron Mueller's third division of
dicotyledonous plants, therefore, it appears that in New South
Wales we have 130 genera, including 602 species. In some of
the families, particularly in Proteaceæ and Candolleaceæ
(Stylideæ), the species are much more numerous in Western
than in Eastern Australia, one whole genus, Dryandra, with
nearly fifty species, being without any representative in the latter
region. In the Rubiaceæ the case is somewhat different, ten only
being western, while thirty-eight are eastern. Of the Composites,
278 species occur in different parts of New South Wales, whilst
in Western Australia only 189 are known. In Campanulaceæ
the species are more equally divided ; but of the genera Diaspasis,
Leschenaultia, Anthotium, Catosperma, and Calogyne, in the
Goodeniaceæ, no species have found their way eastward.

I. DICOTYLEDONEÆ.

(IV.) SYNPETALEÆ HYPOGYNÆ.

The fourth division of dicotyledonous plants are such as usually have connected petals inserted at the bottom of the calyx, stamens affixed to the petals, and fruit laterally free from the calyx. The orders in this division are the same as those described under the Monopetaleæ of Mr. Bentham in the 4th and part of the 5th volumes of the " Flora Australiensis."

1. GENTIANACEÆ.—Limnanthemum, with which Villarsia is now connected, has the following :—L. *geminatum*, L. *exaltatum*, L. *Indicum*, and L. *crenatum*, aquatic or marsh plants, with yellow or white flowers, floating or creeping in habit, and leaves frequently on long petioles. Sebœa *ovata*, with yellow, and Erythrœa *australis*, with pink flowers, are common in most parts of the Colony, and are used medicinally for the bitter principle which pervades them. Gentiana *saxosa* is a small plant, with purplish flowers in compact corymbs, extending from the southern counties to the higher parts of the Australian Alps.

2. The order of LOGANIACEÆ, which is somewhat heterogeneous in its character (differing principally from Apocynaceæ in the nature of the stigma), has the following species of Mitrasacme (small herbs with tetrandrous white flowers, opposite leaves, and two-celled ovary) :—M. *serpillifolia*, M. *paludosa*, M. *pilosa*, M. *alsinoides*, M. *polymorpha*, M. *Indica*, and M. *paradoxa*. Geniostoma *petiolosum* is a large shrub of the order, and is indigenous in Lord Howe's Island (Frag., vol. vii, p. 28). Of the genus Logania, L. *floribunda* and L. *pusilla* are common from Sydney to the Blue Mountains, the one being a shrub several feet in height, and the other a very small plant, only a few inches in length, and procumbent. L. *linifolia* and L. *nuda* are on the Murrumbidgee, Murray, and Darling, the one with small linear leaves, and the other leafless.

3. Of the PLANTAGINEÆ, Plantago *varia*, P. *stellaris*, and P. *Gunnii* are indigenous, the last being an alpine plant. P. *major* (Willd.) and P. *lanceolata* (Willd.) have now established themselves in most parts of the Colony.

4. PRIMULACEÆ.—This order, though widely spread in the Northern Hemisphere, has only three genera in Australia. Lysimachia *salicifolia* and L. *japonica* occur principally in the Northern districts of the Colony ; Samolus *valerandi* and S. *repens* (plants growing in moist or marshy places) are distributed more generally; whilst Anagallis or Pimpernel, A. *arvenis* (Willd.), in its red or blue variety, is common in cultivated ground.

5. MYRSINACEÆ.—Samara *Australiana* is a tall woody climber on the Macleay, Clarence, &c. Myrsine *crassifolia*, M. *platystigma*, and M. *variabilis* are small trees with inconspicuous flowers in

clusters. Ardisia *pseudo-jambosa* is a tree sometimes attaining 30 feet, and having globular purplish berries ; and Ægiceras *majus* is a shrub, growing in marshy places near the sea, and having white sweet-scented flowers. Of these species, Myrsine *varialilis* and Ægiceras *majus* are the only ones found near Sydney.

6. SAPOTACEÆ.—In this order many of the species indigenous in India, Africa, and America are esteemed for their fruits, but in New South Wales the species are not so much in favour. Niemyra *prunifera* (Chrysophyllum of Fl. Aust.) is known from the Bellinger and Clarence rivers ; Amorphospermum *antilogum* (Sersalisia of Fl. Aust.), from the Tweed River ; Sideroxylon (including partly Achras, Sapota, and Sersalisia) has S. *Richardi*, from Illawarra ; S. *Australe*, from Hunter's River to Illawarra ; S. *myrsinoides*, from the Northern districts ; S. *Howeanum*, from Lord Howe's Island ; S. *costatum*, from Norfolk Island ; and the allied Hormogyne *cotinifolia*, from the Northern Ranges.

7. EBENACEÆ, which are generally said to be remarkable for little except the hardness of their woods, are limited to two genera in New South Wales, Diospyros and Maba, the former having D. *mabacea*, D. *Cargillia* (Cargillia *Australis*, R. Br.) and D. *pentamera*, and the latter M. *fasciculosa*. The Australian Ebenads are not valued for their fruits, but some of the Indian and Japanese species (especially D. *kaki*) are esteemed in that respect.

8. The order of STYRACEÆ has only two species in Australia. Symplocos *spicata* and S. *Thwaitesii*, the first being widely spread in the east, and the other probably similar to S. *grandiflora*. They do not extend further south than the Macleay River.

9. The order of JASMINEÆ is represented in the Colony by four species of Jasminum or jasmine, J. *didymum*, J. *lineare*, J. *simplicifolium*, and J. *suavissimum* ; two of Olea, O. *paniculata* and O. *apetala* ; five of Notelæa, N. *ovata*, N. *longifolia*, N. *microcarpa*, N. *ligustrina*, and N. *linearis* ; and one of Mayepea (Chionanthus, Linn.), M. *quadristaminea*, from Lord Howe's Island (Frag., vol. x, p. 89). Some of the native jasmines are sweetly scented, and worthy of cultivation, whilst the fruits of the olives might be utilised for the oil which they contain.

10. APOCYNEÆ.—This order is well characterised by its singular stigma, which is generally expanded at the base into a circular membrane or inverted cup, and is contracted somewhere near the middle. To this stigma the anthers adhere firmly. The fruit varies very much in character, and the habit of the species is equally diverse. Chilocarpus *Australis* is a tall woody climber with large berries ; and Melodinus *Baueri* from Norfolk Island is similar to it in its general appearance. Carissa *Brownii* is a shrub armed with spines, and much branched. The species of Alyxia (A. *buxifolia*, A. *ruscifolia*, A. *squamulosa*, A. *gynopogon*, and A. *Lindii*) are glabrous shrubs with leaves in whorls of three

or four, and flowers in small heads or clusters. Ochrosia *Moorei* is a slender tree with scarlet drupes. Tabernæmontana *orientalis* has dichotomous branches and fruit, with ovoid-falcate carpels, each containing three or four seeds. Alstonia *constricta*, "the Bitter Bark or Cinchona," of the interior is a tall shrub with numerous flowers and follicles 3 or 4 inches in length. Parsonia and Lyonsia are climbing plants, having long follicles. Some of the species are woody, and ascend the highest trees. The following are widely distributed in New South Wales—Parsonia *lanceolata*, P. *velutina*, and P. *ventricosa*, *Lyonsia lilacina*, L. *induplicata*, L. *straminea*, L. *reticulata*, and L. *largiflorens*.

11. ASCLEPIADEÆ.—In this order the species have the anthers and stigma consolidated into a column, and they are for the most part twining plants with a milky juice and follicular fruit. Secamone *elliptica* is a slender twiner. Vincetoxicum *elgans* and V. *carnosum* are similar in habit, but with flowers longer than the leaves. Sarcostemma *australe* is a leafless plant, with flowers having a fleshy double corona. Dœmia *quinquepartita* (Pentatropis, Benth.) has dark purple flowers and linear leaves. Tylophora *grandiflora*, T. *floribunda*, T. *barbata*, T. *Woollsii*, T. *paniculata*, T. *enervis*, and T. *biglandulosa* are herbaceeus plants, with flowers generally dark purple. Marsdenia *flavescens*, M. *tubulosa*, M. *rostrata*, M. *longiloba*, M. *suaveolens*, M. *Leichhardtiana*, and M. *viridiflora* are more robust in habit than those of the preceding genus, and have white or greenish flowers, and in some species large tuberous roots. Hoya *Australis*, which extends from Northern Queensland to the Clarence River, is distinguished by its thick fleshy leaves and waxy flowers. Gomphocarpus *fruticosus*, or the Cape Cotton, is a weed from the Cape of Good Hope, and Asclepias *curassavica* from the West Indies.

12. The order of CONVOLVULACEÆ, which may be briefly characterised by its pentandrous flowers, twining habit, and leafy doubled-up cotyledons, differs considerably in the size and habit of the species, some having showy flowers and climbing amongst trees, and others being small prostrate herbs with inconspicuous flowers. Ipomæa, though a large genus in Australia, has only the following species in New South Wales :—I. *bona nox*, I. *palmata*, I. *peltata*, I. *pescapræ*, and I. *separia*. Convolvulus (which differs from Ipomœa in its filiform style with two stigmatic lobes) has C. *erubescens*, C. *marginatus*, C. *sepium*, and C. *cataractæ*. Polymeria (differing again from Convolvulus in its many-parted style) has P. *longifolia* and P. *calycina*. In Breweria (of which B. *media* occurs in the interior) the style is divided to the base with a capitate stigma on each branch, whilst in Evolvulus (of which E. *linifolius* is common in the Northern districts) the styles are two, filiform, distinct from the base, each divided into two branches with linear stigmas. Cressa *cretica* is a small

diffuse herb of the interior; Dichondra *repens,* a minute creeping plant, pubescent or silky; Wilsonia *humilis,* W. *rotundifolia,* and W. *Backhousii,* prostrate, much-branched herbs, found in saline tracts of the interior, or on marshes near the shore. Cuscuta *Australis* is a leafless threadlike parasite, &c., with nearly sessile flowers in globular clusters. C. *trifolium* (Linn.), which has been introduced with foreign seeds, is sometimes very troublesome on clover and lucerne.

13. The order of SOLANACEÆ, or the Solanum family, is common to the temperate and tropical parts of the world. Most of the species are pentandrous with alternate leaves; they are herbaceous or shrubby, and their fruit is either a two- or a four-celled capsule, or a many-seeded berry. Physalis *Peruviana* (Linn.) (commonly called Cape Gooseberry) is of South American origin; but the smaller species, P. *minima,* is common to New South Wales, Queensland, and North Australia. The large genus Solanum, the anthers of which are characterised in opening by pores, is widely distributed throughout the Colony. The species are—S. *nigrum,* S. *vescum,* S. *aviculare,* S. *simile,* S. *Bauerianum,* S. *verbascifolium,* S. *discolor,* S. *stelligerum,* S. *parvifolium,* S. *ferocissimum,* S. *violaceum,* S. *tetrathecum,* S. *oligacanthum,* S. *esuriale,* S. *chenopodium,* S. *Stuartianum,* S. *densivestitum,* S. *semiarmatum,* S. *armatum,* S. *pungetium,* S. *eremophilum,* S. *campanulatum,* S. *cinereum,* S. *lacunarium,* S. *petrophilum,* S. *ellipticum,* S. *sporadotrichum.* Though the berries of the Solanum are looked upon as narcotic or deleterious, some of them are eaten under the names of " Kangaroo Apples " and " Blackberries." Lycium *Australe* (a plant allied to L. *Chinese* or boxthorn, which may frequently be seen growing wild near gardens) is found in the deserts of the Murray and Darling, and is the only Australian species endemic. Datura *stramonium* (Linn.) and Nicandra *physaloides* (Gaert.) (the seeds of which are decidedly injurious) are plants of foreign origin, but the native tobacco (Nicotiana *suaveolens*) occurs in all the Australian Colonies. N. *glauca* (Grah.) is an introduced plant. Anthocercis, the species of which are generally glandular, pubescent, or hoary, has, for the most part, beyond the Dividing Range, A. *myosotidea,* A. *scabrella,* A. *albicans,* and A. *Eadesii.* Duboisia (a genus placed by some botanists amongst the Scrophularineæ) is represented by D. *myoporoides* near Sydney and D. *Hopwoodii* and D. *Leichhardti* in the northern and western parts of the Colony. The first of these was known for many years to possess deleterious properties (see " Contribution to the Flora of Australia," page 206), and the last is the " Pitury " of the aboriginal natives, which is said to be a stimulant of marvellous power and medicinal efficacy. (See Baron Mueller's remarks and paper read by Dr. Bancroft before the Queensland Philosophical Society, 1877.)

14. SCROPHULARINEÆ.—Mimulus *gracilis*, M. *repens*, and M. *prostratus* are herbs having a five-angled calyx, blue or violet flowers, and opposite leaves. Mazus *pumilio* is a small creeping plant with purple flowers, and sometimes forming large patches in moist places. Stemodia *Morgania* (Morgania of the Fl. Aust.), including the two forms, M. *floribunda* and M. *glabra*, is a herb with opposite or whorled leaves and blue flowers. Bramia *Indica* (Herpestis *monnieria* of the Fl. Aust.) is a small procumbent plant occurring in marshy places near the coast, whilst the genus Gratiola, which has been considered to possess medicinal virtues, has growing in or near water G. *pedunculata*, G. *Peruvania*, and G. *nana*. Artanema *fimbriatum* is a larger plant than any of the preceding. It has racemes of violet flowers, leaves 3 or 4 inches long, and an angular stem. Lindernia *crustacea* and L. *alsinoides* (referred to Vandellia in the Fl. Aust.), Peplidium *humifusum*, Glossostigma *elatinoides*, and G. *Drummondii* are herbs with a diffuse or creeping habit. Limosella *aquatica* and L. *Curdieana* form little tufts in marshy places, the former being identical with the European species. The species of Veronica are mostly small, but the flowers, which are generally blue, may be considered pretty. V. *densifolia* is a densely-tufted plant common to the elevated mountains of Victoria and New South Wales. V. *perfoliata* is considered to mark an auriferous formation, and the rest of the species, V. *Derwentia*, V. *arenaria*, V. *nivea*, V. *gracilis*, V. *calyeina*, V. *plebeia*, V. *notabilis*, V. *serpillifolia*, and V. *peregrina* are small shrubs or herbs. Centranthera *hispida* is a scabrous plant with terminal spikes of flowers, the anthers of which have an awn-like point. Buchnera *urticifolia* and B. *gracilis* are also stiff herbs with dense spikes of flowers. Euphrasia, or "Eye-bright," has three species in the Colony, E. *Brownii*, E. *scabra*, and E. *antarctica*. These have purple or yellow flowers, and delight in moist places. The plants of this order are distinguished by their irregular flowers, stamens usually two or four, and two-celled capsule. Amongst the introduced species may be mentioned Verbascum *Blattaria* (Linn.), Celsia *cretica* (Linn.), Linaria *elatine* (Willd.), and Antirrhinum *orontium* (Mill.)

15. OROBANCHEÆ.—Of this order, Orobanche *cernua* is the only species in Australia, and is common to the Northern Hemisphere and E. Indies. It is a parasitical plant, with flowers of lurid bluish purple, seldom rising to a foot in height, and extending only to the southern part of New South Wales. Mr. Bentham remarks that "its introduction into Australia is not easily accounted for."

16. LENTIBULARINEÆ.—The only genus of the order in New South Wales is Utricularia, which has the following species :—U. *flexuosa*, U. *exoleta*, U. *cyanea*, U. *lateriflora*, U. *dichotoma*,

F

U. *biloba,* and U. *uniflora.* Some of these have floating, and some erect stems. Of the former, two may be regarded as carnivorous plants, as they are said to prey not only on microscopic insects but even small fish.

17. Fieldia *Australis,* a tall climbing shrub, clinging to the trunks of trees on the Blue Mountains and the Southern Ranges, is the only plant of the GESNERACEÆ in New South Wales. Negria *rhabdothamnoides* is peculiar to Lord Howe's Island.

18. The BIGNONIACEÆ are poorly represented in New South Wales, two species only, Tecoma *Australis* and T. *jasminoides,* occurring.

19. Two small plants of the ACANTHACEÆ, Ruellia *Australis* and Eranthemum *variabile,* are widely dispersed throughout the Colony. Hypoestes *floribunda* is known principally from Clarence River; Justicia *procumbens,* J. *Bonneyana,* and J. *eranthemoides* from the northern rivers and the western interior.

20. ASPERIFOLIÆ.—The great majority of species in this order are herbs, but Ehretia *acuminata* is a tree rising to the height of 20 or 30 feet, with panicles of small white flowers and drupes containing several seeds. It is generally deciduous. Heliotropium has the following species in different parts of the interior :— H. *Curassavicum,* H. *Europœum,* H. *asperrimum,* and H. *vestitum*; Halgania *cyanea* and H. *lavandulacea* have showy purple flowers; and the following may be reckoned amongst "the native Forget-me-nots," though strictly speaking Myosotis only should be so named, Pollichia *zeylanica,* Myosotis *Australis,* and M. *suaveolens,* Eritrichum *Australasicum,* Lappula *concava,* Rochelia *Maccoya,* Cynoglossum *latifolium,* C. *suaveolens,* and C. *Australe.* Most of these plants are rough with coarse hairs, and the flowers are in one-sided spikes or racemes, often forked. Echium *violaceum* (Linn.), Anchusa *officinalis* (Linn.), and Lithospermum *arvense* (Linn.), now growing wild in different parts of the Colony, are exotics.

21. The family of the LABIATES or LABIATÆ, which for the most part consists of herbaceous plants, is well defined by a monopetalous bilabiate corolla, on which are inserted the stamens (usually four, but sometimes only two), by opposite leaves replete with volatile oil, and a four-lobed ovary, and with a solitary style rising from the base of the lobes. The species are generally distributed throughout the world, but, with the exception of the genus Prostanthera, which is endemic in Australia, they are neither numerous nor important in New South Wales. Moschosma *polystachya* and Plectranthus *parviflorus* are herbs with pale blue flowers, common to Australia and other parts of the East. Mentha *laxiflora,* M. *grandiflora,* M. *Australis,* M. *gracilis,* and M. *satureoides,* "the Australian Mints," are strongly-scented herbs, closely allied to each other, and possessing medicinal

properties. Lycopus *Australis* is a coarse plant, attaining several feet, having dense whorls of small white flowers, and the two upper stamens reduced to small staminodia. Of Salvia or "Sage," only one species, S. *plebeia*, is indigenous in New South Wales, and that is widely distributed in other regions. Prunella *vulgaris* ("Self-heal") is almost cosmopolitan, whilst the two species of Scutellaria (S. *mollis* and S. *humilis*) or "skull-cap" are endemic. Prostanthera (so called in allusion to the spurs of the anthers) comprises the following species, which are shrubs or under-shrubs with resinous glands strongly scented, one only attaining the size of a tree (P. *lasiantha*) :—P. *prunelloides,* P. *cœrulea,* P. *ovalifolia,* P. *incisa,* P. *rotundifolia,* P. *violacea,* P. *incana,* P. *hirtula,* P. *denticulata,* P. *rugosa,* P. *marifolia,* P. *rhombea,* P. *cuneata,* P. *linearis,* P. *phylicifolia,* P. *empetrifolia,* P. *Behriana,* P. *nivea,* P. *striatiflora,* P. *saxicola,* P. *euphrasioides,* P. *staurophylla,* P. *cryptandroides,* P. *ringens,* P. *Walkeri,* P. *coccinea,* and P. *chlorantha.* According to Baron Mueller, P. *lasiantha* is the largest known labiate, whilst, from experiments made under his direction, many of the species are found to contain valuable oils. Hemigenia *purpurea* and H. *cuneifolia* are shrubs occurring from Port Jackson to the Blue Mountains, but the latter is rare, and frequently mistaken for a Westringia. Of the twenty-two species in Australia these are the only ones on the eastern coast:—Westringia *rosmariniformis,* a robust shrub, found in sandy places near the sea, and W *rigida,* W. *longifolia,* and W. *glabra,* occurring inland. Ajuga *Australis* ("Australian bugle") is found in all the Australian Colonies, and differs but little from a species in Europe. Teucrium *racemosum,* T. *corymbosum,* T. *sessiliflorum,* and T. *argutum* are herbs remarkable for their diversity in habit and inflorescence. Of Labiates, which have become naturalised in the Colony, the following may be enumerated:—Marrubium *vulgare* (Linn.), Stachys *arvensis* (Linn.), Molucella *lævis* (Willd.), Leonitis *leonurus* (R. Br.), and Salvia *verbenacea* (Linn.)

22. VERBENACEÆ.—This order differs from the Labiates in the concrete carpels of the species, their terminal style, and the usual absence of volatile oil in the leaves; but some of them are difficult to distinguish, the two orders being closely allied. Spartothamnus *junceus* is a broom-like shrub, almost leafless, with red succulent drupes. Lippia *nodiflora* is a prostrate herb with purplish flowers strongly scented, and not extending far into New South Wales. Of the genus Verbena, V. *officinalis* is common to all the Australian Colonies. Chloanthes *Stœchadis* and C. *parviflora* are herbaceous plants, remarkable for their greenish flowers and glandular hairs, the leaves being bullato-rugose and decurrent along the stem. Callicarpa *pedunculata* has purplish drupes, and leaves with resinous dots on the under side. Clerodendron *inerme,* C.

tomentosum, and C. *floribundum* are shrubs or small trees with slender corolla-tubes, and stamens exserted, sometimes very long. In C. *tomentosum* the fruiting calyx expands and displays a black shining drupe. Gmelina *Leichhardtii* is a fine timber-tree, known to the colonists as "White Beech." Vitex *trifolia*, V. *Lignum vitæ*, and V. *glabrata* are trees with opposite digitate leaves, and flowers in loose cymes axillary or terminal. Avicennia *officinalis* is a small tree extending along the sea-coast all round the Australian continent, and is frequently called "Mangrove," because it resembles Rhizophora in habit.

Lantana *camara* (Linn.), Verbena *Bonariensis*, and V. *venosa* (G. and H.) are spreading widely in the Colony, and the first two are regarded as troublesome weeds.

23. The order MYOPORINÆ is limited to two genera in New South Wales. Myoporum, which is characterised by its five-parted calyx and corolla, four stamens, and succulent drupes, has near Sydney M. *acuminatum*, a large shrub, and M. *debile*, an almost prostrate plant of a foot or two. M. *tcnuifolium*, M. *montanum*, M. *deserti*, M. *insulare*, M. *viscosum*, M. *humile*, M. *platycarpum*, M. *Bateæ*, and M. *floribundum* are for the most part species of the interior, and M. *obscurum* from Norfolk Island. The genus Eremophila is named most appropriately, as the species are found principally in the arid plains beyond the Dividing Range. E. *Mitchellii* is a small tree, and frequently called "Sandal-wood," the wood being fragrant. Most of the species are mere shrubs, but some of them are remarkable for the beauty of their flowers and succulent drupes. Baron Mueller gives the following for New South Wales:—E. *Bowmanni*, E. *oppositifolia*, E. *Mitchellii*, E. *Sturtii*, E. *Latrobei*, E. *longifolia*, E. *polyclada*, E. *bignoniflora*, E. *Freelingii*, E. *Goodwinii*, E. *Brownii*, E. *Duttonii*, E. *maculata*, E. *latifolia*, E. *alternifolia*, E. *Dalyana*, E. *scoparia*, E. *Macdonnelli*, and E. *divaricata*. Some of the species have been successfully cultivated in Europe, and are admired for the elegance of their flowers.

24. Of the ERICACEÆ, or Heathworts, which are widely distributed throughout the temperate and colder regions of the world, there are very few in Australia, and only one of these occurs on the snowy summits of the highest mountains of New South Wales. Gaultheria *hispida* is common to Victoria and Tasmania, as well as this Colony, and it is a spreading shrub of a few feet in height, with white flowers in dense racemes. This plant is distinguished from the Epacrids by the terminal openings of the anther-cells, each with two erect awns.

25. The species of the EPACRIDS (Epacrideæ), which for the most part are limited to Australasia, are amongst the most admired of native plants, and differ from the preceding family

in the anthers opening by longitudinal slits, and in the simple or forked nerves of the leaves, which resemble those of endogens. The species of Epacris are favourites with floriculturists, and the double flowers of E. *purpurascens*, E. *impressa*, and E. *microphylla* have rendered them objects of peculiar interest, whilst the general observer looks upon the order in Australia as a natural substitute for the Heaths of the Cape of Good Hope. The Epacrideæ naturally divide themselves into two groups, viz., those with drupaceous and those with capsular fruit.

(1.) Tribe with drupaceous fruit :—

In his systematic list, the Baron, following the principles enunciated by some of our early botanists, has amalgamated in the genus Styphelia the various genera of dupraceous Epacrids (Astroloma, Lissanthe, Leucopogon, Acrotriche, Monotoca, &c.), believing that they should be regarded simply as so many sections of that one genus. As this arrangement, however, has not been adopted by Mr. Bentham, nor is agreeable to those who do not like to give up Brown's time-honored names, it may be advisable to distinguish the species of Styphelia otherwise designated. In the Baron's list S. *adscendens*, S. *longifolia*, S. *lœta*, S. *triflora*, S. *viridis*, and S. *tubiflora* are the generally recognised species of Styphelia, and known by the appellation of "Five Corners." S. *humifusa*, S. *Sonderi*, and S. *pinifolia* are described in the "Flora Australiensis" under the genus Astroloma or "Groundberry." S. *procumbens* and S. *urceolata*, shrubs of similar habit, have hitherto been referred to Melichrus, whilst S. *sapida* and S. *strigosa* (the former a good-sized shrub, with edible berries, and the latter much smaller, with pretty little white flowers) are species of Brown's genus Lissanthe. By far the greater number of the Baron's Styphelia are species of Leucopogon, a genus so called from the bearded segments of the corolla. They are for the most part small heath-like shrubs, varying in habit and size, having flowers very similar in character, and fruits which in a few species may be regarded as edible. These plants are S. *amplexicaulis*, S. *lanceolata*, S. *Richei*, S. *Australis*, S. *collina*, S. *microphylla*, S. *virgata*, S. *montana*, S. *Macraei*, S. *linifolia*, S. *pluriloculata*, S. *pleiosperma*, S. *attenuata*, S. *conferta*, S. *mutica*, S. *ericoides*, S. *margarodes*, S. *esquamata*, S. *cordifolia*, S. *biflora*, S. *setigera*, S. *exolasia*, S. *Fraseri*, S. *juniperina*, S. *rufa*, S. *deformis*, S. *appressa*, and S. *neo-anglica*. These species of Leucopogon are widely diffused throughout the Colony, some occurring near the coast, some on the mountains, and a few beyond the Dividing Range. S. *divaricata*, S. *aggregata*, and S. *serrulata* belong to Acrotriche ; and S. *elliptica*, S. *scoparia*, and S. *ledifolia* to Monotoca (so called because the ovary is usually one-celled). S. *daphnoides* and S. *ericoides* are of the genus or section Brachyloma, small shrubs similar in habit to

Leucopogon. S. *laurina* and S. *pumila* are species of Trocho-
carpa, the former being a middle-sized tree with useful wood,
and the latter a diffuse or prostrate shrub known only from the
mountains of the south. By the names now enumerated it will
be seen that the Baron makes of them fifty species of Styphelia, whilst in the Flora they are arranged in nine genera.

(2.) Tribe with capsular fruit:—
In this tribe the ovules are several in each cell of the ovary,
the style is inserted in a central depression, and the capsule
opens in the carpels dehiscing through their backs. The species
of the genus Epacris, which is almost limited to Australia and
New Zealand, are in New South Wales:—E. *longiflora*, E. *reclinata*,
E. *impressa*, E. *sparsa*, E. *petrophila*, E. *rigida*, E. *coriacea*, E. *crassifolia*, E. *robusta*, E. *obtusifolia*, E. *paludosa*, E. *Calvertiana*, E.
heteronema, E. *serpillifolia*, E. *microphylla*, E. *apiculata*, E. *pulchella*, and E. *purpurascens*. Lysinema *pungens* is separated by
the Baron from the Western genus under the name Woollsia, as
being peculiarly an Eastern species with the habit of an Epacris,
and with the filaments not always distinctly free. Ponceletia
sprengelioides, P. *montana*, and Sprengelia *incarnata* are found in
marshy places or on wet rocks on the mountains ; whilst Draco-
phyllum *secundum*, which is known only in New South Wales,
occurs in moist rocky places from Port Jackson to the Blue Mountains, and at Illawarra. D. *Fitzgeraldi* (so named in honor of Mr. R.
D. Fitzgerald, F.L.S., who discovered it in Lord Howe's Island)
is a large tree attaining sometimes a height of 40 feet, with
leaves nearly a foot long. A full description and figure of this
remarkable species are given in the Frag., vol. vii, p. 27, by
Baron Mueller.

I. DICOTYLEDONEÆ.

(V.) APETALEÆ GYMNOSPERMEÆ.

In the last division of the Baron's, viz., Apetaleæ Gymnospermeæ,
the Coniferæ and Cycadeæ (which in the Flora are associated with
the Monochlamydeæ in the 6th vol.) are placed separately, being
apetalous, and having, as Lindley observes, "nearly an equal
relation to flowering and flowerless plants," showing "a plain
transition from the highest forms of organisation to the lowest."
1. CONIFERÆ.—In this order Araucaria *Cunninghami*, "the
Moreton Bay pine," and A. *excelsa*, "the Norfolk Island pine,"
are the most important, in consideration of their great size,
ornamental character, and industrial properties. Of the genus
Callitris or Frenela the following species belong to New South
Wales, but only one species occurs near Sydney:—C. *Macleayana*,
C. *Parlatorei*, C. *verrucosa*, C. *columellaris*, C. *Muelleri*, C. *cupressi-
formis*, and C. *calcarata*. Some of these grow to a considerable

size, and afford almost the only timber in certain parts of the interior. Pherosphæra *Fitzgeraldi* is a highly interesting shrub found at Katoomba by Mr. R. D. Fitzgerald, and attaining only a few feet in height. It seems to connect the alpine flora of New South Wales with that of New Zealand. Nageia (the Podocarpus of the Flora) is represented by N. *elata*, N. *spinulosa*, and N. *alpina*. The first of these is a tree of considerable size, the second a diffuse kind of shrub with edible fruits, and the third a straggling alpine species on the highest parts of our southern mountains.

2. CYCADACEÆ.—The true Cycas does not extend to New South Wales, but of the genus Encephalartos (including Macrozamia) the Baron reckons for this Colony E. *Pauli-Gulielmi*, E. *tridentatus*, E. *spiralis*, and E. *Denisonii*. To these Mr. C. Moore, F.L.S., has added the following species :—M. *cylindrica*, M. *secunda*, M. *Fawcetti*, M. *flexuosa*, and M. *heteronema*. ("A Census of the Plants of New South Wales," by Charles Moore, F.L.S., &c., p. 66.) M. *corallipes* (regarded by Mr. Bentham as a variety of M. *spiralis**) is also stated to be distinct. With this order the first great division of the Dicotyledoneæ ends, and, so far as yet ascertained, the number of species is as follows :—

		Gen.	Spec.
I.	Choripetaleæ Hypogynæ	236	710
II.	„ Perigynæ ...	128	623
III.	Synpetaleæ perigynæ ...	131	602
IV.	„ Hypogynæ ...	119	391
V.	Apetaleæ Gymnospermeæ	5	23
		619	2,349

II. MONOCOTYLEDONEÆ.

(I.) CALYCEÆ PERIGYNÆ.

The orders of monocotyledonous plants with the stamens inserted on the tube of the calyx, and at a distance from the base of the ovaries, the fruit with few exceptions being laterally adnate to the tube of the calyx.

1. ORCHIDEÆ, or Orchids, are amongst the most admired of monocotyledonous plants. The peculiarity of their structure in the consolidation of stamens and pistil into one mass ; the unusual figure of their flowers, sometimes resembling an insect or reptile, and sometimes a helmet with the visor up, or some fantastic shape ; and the general distribution of the species, occurring as

---

* The seeds of certain species of Encephalartos, after having been subjected to a process of "pounding, maceration, and desiccation" (Thozet), are by the blacks utilised for food. In a raw state the seeds are deemed poisonous. It is said that a preparation of them is the antidote for snake-poison of which some years since a person named Underwood spoke confidently, but who subsequently lost his life from the bite of a snake.

they do in all parts of the world, excepting in the coldest regions —all these circumstances combine together to make the order a favourite one amongst observers and floriculturists. In Australia the known species are between 200 and 300 (255 being enumerated by the Baron, and others being added by Mr. R. D. Fitzgerald, F.L.S., in his beautiful work on Australian orchids), so that next to the Cyperaceæ and Gramineæ the species are more numerous than in any other order of monocotyledoneæ. According to Mr. Bentham, two-thirds of the species are essentially Australian, whilst four of the genera are represented by single or very few species in the Indian Archipelago, and eleven have New Zealand congeners, sometimes identical in species. The genera have been arranged as follows by Baron Mueller :—

1. Sturmia (Liparis of the Fl. Aust.) has in New South Wales two species, L. *reflexa* and I,. *cœlogynoides*, small terrestrial or epiphytal plants, with greenish-yellow flowers and shortly-creeping roots.

2. Oberonia *iridifolia* and O. *palmicola* are diminutive plants similar in habit to Sturmia, but with very small flowers in dense racemes, and not extending further south than the Hastings.

3. Dendrobium.—This genus, which includes the so-called "Rock Lily" of the colonists and other species creeping over rocks or growing on trees near Sydney, has the following in different parts of the Colony:—D. *speciosum*, D. *falconirostre*, D. *tetragonum*, D. *æmulum*, D. *Kingianum*, D. *brachypus*, D. *gracilicaule*, D. *Moorei*, D. *monophyllum*, D. *cucumerinum*, D. *pugioniforme*, D. *linguiforme*, D. *teretifolium*, D. *Fairfaxii*, D. *striolatum*, D. *Mortii*, D. *macropus*, and D. *Beckleri*.

4. Bulbophyllum is nearly allied to the last, but the species are mere herbs, with a creeping root and leaves on small pseudo-bulbs. They are B. *Shepherdi*, B. *aurantiacum*, B. *argyropus*, B. *exiguum*, B. *minutissimum*, the smallest of Australian orchids, and B. *Elisæ*.

5. In Sarcochilus, the labellum has not any spur at its base, but its terminal lobe has a solid fleshy protuberance. S. *erectus*, S. *tridentatus*, S. *Beckleri*, S. *divitiflorus*, S. *falcatus*, S. *Fitzgeraldi*, S. *Ceciliæ*, S. *olivaceus*, S. *parviflorus*, and S. *Hillii* are widely distributed in the eastern parts of the Colony, some occurring in the gullies of the Blue Mountains. S. *Fitzgeraldi* (named in honor of its discoverer) is one of the rarest and prettiest of Australian orchids, and S. *parviflorus* is remarkable as being the most southern species of the genus.

6. Ornithochilus *Hillii* (Saccolabium of the Fl. Aust.) is a small epiphytal species, not extending far south.

7. Dipodium *punctatum* is a tall leafless plant with spotted flowers, generally pink, but varying in colour.

8. Cymbidium *canaliculatum* and C. *albuciflorum* are northern species with loose racemes of greenish-yellow flowers and long narrow leaves. C. *suave* is found as far south as Illawarra. C. *canaliculatum* was noticed by Sir T. Mitchell (Tropical Australia, p. 378) in forests near the Balonne, and M. Thozet states that the blacks used the bulbous stems as food.

9. Phajus *grandiflorus* (Phaius of the Fl. Aust.) has large showy flowers and leaves, the leafy stems thickening into pseudo-bulbs. It has been collected on the Macleay, Tweed, and Richmond Rivers.

10. Calanthe *veratrifolia* differs but little from the Indian species, and extends from Queensland to the Blue Mountains and Illawarra. It is a terrestrial plant with large racemes of white flowers and leaves 2 or 3 feet long.

11. Galeola *cassythoides* and G. *foliata*, the largest of Australian orchids yet known, are leafless plants with flexuose branches and a climbing habit. The one occurs near Sydney, but the other, which ranges from Queensland to the Clarence, is the larger species, with pendulous panicles of flowers, broader and more branched than those of the preceding.

12. Gastrodia *sesamoides* is a parasitical herb with leafless stems and flowers in a terminal raceme.

13. Epipogum *nutans* is also leafless, known only from the Tweed River, and similar in all respects to the Indian species.

14. Spiranthes *Australis* is common to all the Australian Colonies, Tasmania. New Zealand, a great part of Asia, and some parts of Europe.

15. Thelymitra is a genus of terrestrial plants, the species of which have blue, purple, or red flowers, some showy and interesting. The species near Port Jackson are T. *ixioides*, T. *circumsepta*, T. *aristata*, T. *longifolia*, T. *carnea*, and T. *venosa*.

16. Diuris, which is characterised by its narrow lateral sepals, is similar to the last in habit, but has white, yellow, or purplish flowers. The species, some of which are very common in the spring, are D. *alba*, D. *punctata*, D. *secundiflora*, D. *aurea*, D. *maculata*, D. *pedunculata*, D. *pallens*, D. *abbreviata*, D. *sulphurea*, D. *æqualis*, and D. *dendrobioides*. There is a small variety of D. *punctata* which flowers rather later in the season.

17. Orthoceras *strictum* in some respects resembles Diuris, and is common to New Zealand and Australia.

18. Calochilus *campestris*, C. *Robertsoni*, and C. *paludosa*, common to several of the Australian Colonies, are terrestrial orchids, remarkable for their beards or densely-fringed labellum.

19. Cryptostylis *longifolia* and C. *erecta*, plants with rather large flowers, green with a brown-red or purple labellum, occur near Sydney and Parramatta, but C. *leptochila* is known only from the Blue Mountains.

20. Prasophyllum varies very much in size and colour. P. *Australe*, P. *elatum*, P. *brevilabre*, P. *flavum*, P. *patens*, and P. *fuscum* may be reckoned amongst our tallest orchids, especially the first two ; whilst P. *striatum*, P. *nigricans*, P. *rufum*, P. *fimbriatum*, and P. *Woollsii* are small plants rising only to the height of 6 or 8 inches.

21. Microtis *porrifolia* (including M. *parviflora*) has green flowers in a terminal spike, and sometimes exceeds a foot in height.

22. The species of Corysanthes, C. *unguiculata*, C. *pruinosa*, C. *fimbriata*, and C. *bicalcarata*, are dwarf herbs with a single cordate or reniform leaf and dark-coloured helmet-shaped flowers.

23. Pterostylis.—The species vary from a few inches to a foot, have green helmet-shaped flowers, leaves either cauline or radical, and for the most part a winged column. From Sydney to the Blue Mountains the following occur (one or two extending far inland) :—P. *ophioglossa*, P. *concinna*, P. *curta*, P. *acuminata*, P. *nutans*, P. *pedaglossa*, P. *pedunculata*, P. *grandiflora*, P. *truncata*, P. *reflexa*, P. *præcox*, P. *obtusa*, P. *parviflora*, P. *barbata*, P. *mutica*, P. *rufa*, P. *Daintreyana*, P. *longifolia*, P. *cucullata*, and P. *nana*. Some authors make more species, but the Baron unites under P. *rufa* several forms differing principally in the length of the sepal points, and perhaps distinct species.

24. Caleya *major* and C. *minor* are small plants remarkable for the irritability of their labellum.

25. Drakæa is allied to Caleya. D. *irritabilis* is found in Queensland and New South Wales. The western species are known by the name of "Hammer Orchids."

26. Acianthus *caudatus*, A. *fornicatus*, and A. *exsertus* are common in some parts of the Colony, and one species of the genus extends to New Zealand.

27. Cyrtostylis *reniformis* is a delicate little species with one orbicular cordate leaf and a raceme of a few pale-red flowers.

28. Lyperanthus *nigricans* seldom exceeds a few inches, and in drying turns black. L. *Burnettii* (B. *cuneata* of Fl. Aust.) is a Tasmanian species, but Mr. R. D. Fitzgerald has found it on the Blue Mountains.

29. Eriochilus *autumnalis* is a glandular pubescent herb, and, as its name implies, flowers late in the season.

30. The species of Caladenia are pretty little plants of different colours, varying considerably in size, with rows of glands on the labellum, a solitary linear leaf, and in some species with long sepals, which have gained for the flowers the name of "Spider Orchids." The Baron gives for New South Wales, C. *Nortoni*, C. *Patersoni*, C. *latifolia*, C. *suaveolens*, C. *carnea* (including C. *alba*), C. *congesta*, C. *cœrulea*, and C. *deformis*. From the elegant

figures in Mr. Fitzgerald's "Australian Orchids" it will be seen that there is reason for supposing that some of the forms referred to *C. Patersoni* are distinct species.

31. Of Chiloglottis, *C. diphylla* was figured by F. Bauer in the early part of the present century, as known from Port Jackson. To this Mr. Fitzgerald has recently added *C. trapeziformis* and *C. formicifera*. *C. Gunnii* is common to Tasmania, Victoria, and New South Wales, and *C. trilabra* has been added by Mr. Fitzgerald.

32. Glossodia *major* is a pretty species with flowers of a lilac or bluish colour, and an oblong or lanceolate leaf, sweetly scented. *G. minor* is a much smaller plant with darker flowers. It appears that in New South Wales the orchids are referred to thirty-two genera comprising about 130 species. Whether regarded for the singularity of their flowers or the various ways in which their fertilisation is accomplished, the species are highly interesting. These beautiful plants are most abundant near the coast, whilst beyond the Dividing Range they are very limited. Sir T. Mitchell collected only two (Cymbidium *canaliculatum* and Pterostylis *Mitchellii*) in his expeditions, and few with the exception of some species of Diuris and Caladenia extend to the Castlereagh and Macquarie. Pterostylis *rufa* in its various forms may be found from Port Jackson to the Darling.

2. The SCITAMINEÆ, or Gingerworts, are poorly represented in New South Wales, only one, Alpinia *cœrulea*, extending as far south as Hunter's River. It has leafy stems several feet in height, tuberous roots, and blue racemose flowers.

3. The Iris family has in the southern parts of the Colony Diplarrhena *morœa*, with large white flowers sometimes tinged with violet and yellow. Patersonia, with its three species, P. *glauca*, P. *sericea*, and P. *glabrata*, all with blue or purple flowers, is more widely distributed. Iris *Robinsoniana*, the largest species of the genus, is the "Wedding Flower" of Lord Howe's Island. Sisyrinchium (the Libertia of the Fl. Aust.) has S. *paniculatum* in Victoria and New South Wales, and S. *pulchellum*, common to Tasmania, Victoria, and New South Wales. These plants, which occur in moist shady places or in gullies on the mountains, have white flowers clustered in the axils of sheathing bracts and leaves mostly radical. S. *micranthum* (Cav.) (a very small species with yellow flowers) and S. *Bermudianum* (Linn.) with bluish flowers, are introduced plants, but the former is now widely scattered through Queensland and New South Wales. Trichonema *bulbocodium* (H. K.) and Sparaxis *tricolor* (H. K.) may also be reckoned amongst the naturalised Irideæ.

4. Burmannia *disticha* (BURMANNIACEÆ), the only species of the order in New South Wales, is a plant of a foot or two in

height, with short radical leaves and greenish flowers. It is rare near Sydney, but more common in swampy places to the north.

5. DIOSCORIDEÆ.—This order is placed by Lindley amongst what he terms Dictyogens, because the species differ from Endogens generally in having broad net-veined foliage, thus showing a transition class partaking of the texture of Exogens. Of these plants, usually called yams, D. *transversa* is a glabrous twiner, with large tuberous roots, opposite hastate or cordate leaves, and winged seeds. It is not known further south than Hunter's River. Petermannia *cirrhosa* is similar in habit to the last, but more like some species of Smilax than Dioscorea.

6. HYDROCHARIDEÆ.—This order consists of aquatic plants ; Hydrocharis *morsus ranæ* occurs in Europe and Asia, as well as in Australia. Halophila *ovata* (Caulinia *ovalis*, R. Br.) has creeping stems rooting under water, the flowers being enclosed in involucres. Ottelia *ovalifolia* is common in lagoons and ponds, the radical leaves in tufts at the bottom of the water, and the floating ones oblong ovate, the flowers white and enclosed in a tubular two-lobed spathe. Vallisneria *spiralis* is a submerged plant, with long narrow leaves and dioecious flowers, abundant in the lagoons near the Hawkesbury. The female flowers are on spirally-coiled filiform peduncles, which, at a certain season, unfold so as to convey the flowers to the surface. The male flowers break off from their peduncles at the same time, and rising to the surface fecundate the females. Hydrilla *verticillata* is known principally from the Richmond River, and is a submerged herb, with dioecious flowers, cylindrical fruit, and short four to eight-whorled leaves.

7. The AMARYLLIS family is more abundant in Western than in Eastern Australia, Phlebocarya, Trinobanthes, Conostylis, and Anigozanthus being limited to the former. Hœmodorum *planifolium* and H. *teretifolium* are remarkable for their dark flowers and red spongy roots. Curculigo *ensifolia* has yellow flowers in short spikes, long fibrous roots, and radical leaves. Hypoxis *hygrometrica* and H. *glabella* are small herbs with yellow flowers. Doryanthes, "the gigantic lily" of the colonists, was one of the first genera figured by Bauer. D. *excelsa* rises to the height of 18 feet, has red flowers in a dense head of a foot in diameter, and very numerous leaves from the root a foot long or more. There is a white variety of this species. D. *Palmeri* and D. *Larkinii* are from the northern parts of the Colony, and though not so tall as the preceding species, are highly interesting plants. Crinum *flaccidum* is a plant with white flowers in umbels, scape 2 feet high, and elongated leaves. This species grows on the Murray and the Darling, but C. *pedunculatum* has a more extensive range, being found occasionally from Port Jackson to the Castlereagh. Eurycles *Cunninghamii* is a fine plant with numerous white flowers, arranged in umbels, and with leaves of an ovate

shape on long petioles. Calostemma *purpureum* is a bulbous plant with linear leaves, flowers white, pink, or purple, in umbels, and a corona reaching to about half the length of the segments. This species, as well as C. *luteum* (which is similar in habit, but has larger flowers, yellow or white), is found on many of the rivers beyond the Dividing Range, and also westward of the Darling. It is the C. *candidum* of Lindley, of which Sir Thomas Mitchell, when travelling near the Namoi, says :—" I found a flowery desert, the richest part of the adjacent country being quite covered with a fragrant white amaryllis in full bloom." Of C. *carneum*, a variety of C. *purpureum*, he adds:—" I found a few bulbs of a pink-coloured amaryllis, which grew on the summit of the Goulburn Range." He had previously remarked: " On the plains (near the Lachlan) grew in great abundance that beautiful species of lily found in the expedition of 1831, and already mentioned under the name of Calostemma *candidum*, also the C. *luteum* of Ker, with yellow flowers." (Exp., vol ii, pp. 30, 39.) Zephyranthes *atamasco* (Herb.) may be found growing wild near the coast in moist places.

II. MONOCOTYLEDONEÆ.

(II.) CALYCEÆ HYPOGYNÆ.

Those orders of monocotyledonous plants with stamens inserted on the bottom of the calyx and at the base of the ovary ; the fruit, with some few exceptions, laterally free from the calyx.

8. The LILIACEÆ, as now arranged, include the rush-like Xerotes and the truly Australian Xanthorrhœa, or " grass-tree," and they number for all Australia over 160 species. Some of them extend to Tasmania and New Zealand and the Isles of the Pacific, but twenty-four of the genera are strictly endemic.

Phormium *tenax*, or New Zealand flax, which is remarkable for its large fibrous leaves, is recorded very early as a Norfolk Island plant.

Smilax *glycyphylla*, or Australian sarsaparilla, has obtained a reputation for its medicinal properties. S. *Australis* is a climber of considerable size, armed with troublesome prickles. S. *purpurata* is likewise recorded from New South Wales. Belonging to the same tribe, and having also leaves with distant primary veins and transverse veinlets, are Rhipogonum *album*, R. *discolor*, R. *Fawcettianum*, and R. *Elseyanum*, the first of which only occurs near Sydney. Flagellaria *Indica* is a tall glabrous climber, ascending sometimes to the tops of trees. It is a species found also in the tropical parts of Asia and Africa. This plant, as well as the perennial Drymophila *cyanocarpa* and D. *Moorei*, is not recorded to the south of Port Jackson, though D. *cyanocarpa*

appears again on the mountains of Victoria and Tasmania.
Dianella *Tasmanica*, D. *longifolia*, D. *revoluta*, and D. *cærulea*
are plants with crowded distichous leaves and blue flowers.
They are abundant in many parts of the Colony ; but the species
are difficult to define, as they approach each other very closely.
Eustrephus *Brownii* and Geitonoplesium *cymosum* are climbing or
straggling plants with orange-coloured and blue berries. Cordyline
terminalis, C. *stricta*, and C. *Baueri* are shrubs similar to Dracæna,
having the branches marked by the annual scars of the fallen
leaves. They are frequent on some of our northern rivers.
Blandfordia *grandiflora*, B. *nobilis*, and B. *flammea* are tall plants
growing generally in swampy places, and conspicuous for their
red and yellow pendulous flowers. Astelia *alpina* is a densely-
tufted herb clothed with silky hairs, common to the higher
mountains of Victoria and New South Wales. Wurmbea *dioica*
(Anguillaria *dioica*, R. Br.) is a small plant with white flowers,
generally diœcious, and occurring in various forms in all
the Australian Colonies. Schelhammera *undulata* and Kreyssigia
Cunninghami are by some referred to the same genus—the one
small, and found in moist shady places from Port Jackson to the
Blue Mountains, and the other on the banks of northern rivers.
Burchardia *umbellata* has white flowers in a terminal umbel.
Bulbine *bulbosa* and B. *semibarbata* have radical leaves and
racemes of yellow flowers with bearded filaments. Thysanotus
tuberosus, T. *Baueri*, T. *junceus*, and T. *Patersoni* are the so-
called fringed violets of the colonists. The last species differs
from the rest in having wiry twining stems. Cæsia *vittata* has
lilac flowers with a dark stripe in each petal. C. *parviflora* has
small white flowers. Chamæscilla *corymbosa* is a perennial with
numerous blue flowers, more common in Victoria and Tasmania
than in New South Wales. Corynotheca *lateriflora* (Cæsia
lateriflora, R. Br.) belongs to the interior and the sandy ridges
of the Murray. Tricoryne *elatior* and T. *simplex* have yellow
flowers in umbels, and the fruit divided to the base into three
one-seeded nutlets. Stypandra *glauca* and S. *cæspitosa* have blue
flowers, the stamens being filiform and flexuose, with a dense
woolly tuft under the anther. Arthropodium *paniculatum*, A.
minus, A. *strictum*, and A. *laxum* (Dichopogon *strictus*, R. Br.,
and D. *Siberianus*, Kunth.) have grass-like leaves, and white
or purplish flowers. Herpolirion *Novæ-Zealandiæ* is limited to
the higher ranges of Victoria, Tasmania, and New South Wales,
being a dwarf stemless plant, with solitary flowers almost sessile
within the leaves. Sowerbæa *juncea* has globular umbels of pink
flowers, with a few filiform leaves at the base. It delights in
moist and boggy places. Allania *Cunninghami* is a perennial,
known only from the Blue Mountains. It has diffuse stems
covered with linear leaves, and flowers in globular umbels.

Bartlingia *gracilis* and B. *sessiliflora* (Laxmannia, R. Br.) are
herbs with narrow linear leaves and small white flowers, with
imbricate scarious bracts. The genus Xerotes, formerly connected with Juncaceæ, has diœcious flowers and harsh tufted or
radical leaves. The following species are found from the coast
to the interior :—X. *longifolia*, X. *sororia*, X. *effusa*, X. *micrantha*,
X. *Thunbergii*, X. *glauca*, X. *elongata*, X. *rupestris*, X. *flexifolia*,
and X. *leucocephala*.

The grass-trees, some small and others assuming an arbor-
escent size, and forming a peculiar feature in Australian vegetation, are limited to this continent. Two species are western.
Those of New South Wales are X. *macronema*, X. *hastilis*, X.
arborea, X. *bracteata*, and X. *minor*.

9. PALMÆ.—Calamus *Muelleri*, from the Clarence and Richmond rivers, is a "bush lawyer," armed with prickles or bristles,
and differing in that respect from C. *Australis*, which has the
rachis underneath armed with recurved prickles. Bacularia *monostachya* (Areca. Mart., and also Kentia, F. v. M.) is the "walking-
stick palm" of the Northern Districts. Kentia *Belmoreana*, K.
Fosteriana, K. *Baueri*, and K. *Canterburyana* are, with the exception of K. *Baueri* from Norfolk Island, limited, so far as known,
to Lord Howe's Island, where they are called "thatch palm"
and "umbrella palm." Clinostigma *Mooreanum* is also from
Lord Howe's Island. Ptychosperma *Cunninghamii* (Seaforthia
elegans, Hook.) is the "Bangalow" of Illawarra, &c., and Livistona
Australis, "the cabbage-tree palm," occurring near the coast, on
the Blue Mountains, and at Illawarra.

10. Of the PANDANACÆ, or "Screw Pines," Pandanus *pedun-
culatus* grows on the Richmond and Hastings rivers, and P.
Forsteri on Lord Howe's Island. P. *Moorei* belongs also to New
South Wales. They have the aspect of gigantic Bromelias, with
the stems stoloniferous at the base. The side of the seeds adhering to the rachis is eaten by the blacks. Freycinettia *Baueriana*
is indigenous at Norfolk Island.

11. AROIDEÆ.—The common Arum with dark purple spathes,
Typhonium *Brownii*, occurs in moist shady places, from Port
Jackson to the Blue Mountains, and Colocasia *macrorrhiza* from
the Hastings to Kiama. These plants are very acrid, but the
tuberous roots after certain preparation are eaten by the blacks.
Gymnostachys *anceps* has tuberous roots, very long leaves, and
spikes of flowers clustered in the axils of leafy bracts. Pothos
Loureiri differs from the preceding plants in habit, clinging to
the stems of trees, and having leaves with phyllodineous petioles
5 or 6 inches long.

12. TYPHACEÆ.—Typha *angustifolia*, commonly called "Bull-
rush," and Sparganium *angustifolium* are aquatic or marsh plants.

13. LEMNACEÆ.—This order consists of little floating plants without distinct stems or real leaves. Lemna *trisulca*, L. *minor*, L. *oligorrhiza*, and L. *polyrrhiza* emit fibres or roots; but the minute Wolffia *arrhiza* has simple fronds.

14. FLUVIALES.—The plants of this order are abundant in rivers, lagoons, and ponds. Triglochin *centrocarpa*, T. *striata*, T. *procera*, and T. *Maundii* have erect scapes and radical leaves. The "Pondweeds" common in rivers and lagoons are Potamogeton *natans*, P. *perfoliatus*, P. *crispus*, P. *obtusifolius*, P. *javanicus*, and P. *acutifolius*. Ruppia *maritima* and Zostera *nana* are frequent in estuaries and salt lagoons. The latter was much used in the early days of the Colony for stuffing beds. Posidonia *Australis* is a marine submerged plant with larger leaves than those of Zostera. Naias *tenuifolia* is a submerged freshwater species common in the rivers of New South Wales. Aponogeton *elongatus* extends from Queensland to the Clarence and Richmond rivers. It has tuberous roots, which are baked and eaten by the blacks.

15. Of the ALISMACEÆ, the cosmopolitan Alisma *plantago* is widely distributed in lagoons and marshy places. Damasonium *Australe* (Actinocarpus, R. Br.) is a marsh herb similar to the last in habit.

16. PHILYDREÆ.—Philydrum *lanuginosum* is a tall marsh plant plentiful in New South Wales. Helmholtzia *acorifolia* is also a large one, with a dense terminal panicle of flowers. It is on the Richmond River.

17. COMMELINEÆ.—The plants of this order in Australia are erect or creeping in habit, with blue or white flowers, and leaves parallel veined with sheathing bracts. Commelina has C. *ensifolia* and C. *cyanea*, and C. *africana* (Willd.) is an introduced plant; Aneilema, A. *acuminatum*, A. *biflorum*, and A. *gramineum*; and Pollia, P. *cyanococca*. In this sub-order the perianth segments almost represent a calyx and corolla, the three outer ones being smaller than the three inner ones, the latter delicate and petal-like.

18. The order of XYRIDEÆ is represented by one genus, Xyris, and three species, X. *complanata*, X. *operculata*, and X. *gracilis*. The flowers are yellow, solitary and sessile within imbricate, glume-like scales, forming a terminal head.

19. The rush family, JUNCACEÆ, have the following species in New South Wales :—Luzula *campestris* and L. *longiflora*, the latter from Lord Howe's Island, and of Juncus, a genus spread over almost all known parts of the world, J. *planifolius*, J. *cœspititius*, J. *falcatus*, J. *bufonius*, J. *homalocaulis*, J. *Brownii*, J. *communis*, J. *vaginatus*, J. *pauciflorus*, J. *pallidus*, J. *maritimus*, J. *prismatocarpus*, and J. *pusillus*.

20. ERIOCAULEÆ.—Eriocaulon *Australe*, E. *Smithii*, and E.
electrospermum are the only species of Eriocauleæ in New South
Wales. They are herbs with the leaves in radical tufts.
21. RESTIACEÆ.—The species of this order resemble those of
the rush and sedge kind, but they are readily distinguished from
both by the pendulous ovules and seeds. As the male and female
flowers differ very much in appearance there is sometimes much
difficulty in distinguishing the species. Trithuria *submersa* is a
dwarf tufted annual, with filiform radical leaves, and growing
under water. Aphelia *gracilis* and A. *pumilio* are also small
tufted plants from the southern parts of the Colony. Centrolepis,
in the species C. *polygyna*, C. *glabra*, C. *Drummondii*, C. *fascicularis*, and C. *strigosa*, is more widely distributed. They are
small uninteresting annuals. Lepyrodia *scariosa*, L. *Muelleri*,
L. *anarthria*, L. *gracilis*, and L. *interrupta* are rush-like plants,
with sheathing scales. Restio has the following species :—R.
fastigiatus, R. *dimorphus*, R. *Australis*, R. *gracilis*, R. *complanatus*,
and R. *tetraphyllus*. Some of them attain a height of several
feet, and are common in swamps near Sydney. Calostrophus.
lateriflorus and C. *fastigiatus* (Hypolœna, R. Br.) have branched
and flexuose stems, and the glumes slightly woolly. Leptocarpus
tenax and L. *Brownii* are plants with simple branched stems,
leafless except the sheathing scales, which are closely appressed.
The order, Mr. Bentham remarks, is almost limited to extra-
tropical South Africa, Australia, and New Zealand, but the
Australian species are nearly all endemic.

II. MONOCOTYLEDONEÆ.

(III.) ACALYCEÆ HYPOGYNÆ.

Plants without a calyx, with stamens (in bisexual flowers)
inserted at the base of the ovary, and with fruit adnate to or free
from its glumaceous bract; exceptions rare.
22. The CYPERACEÆ, or Sedges, numbering for all Australia
nearly 400 species, constitute a large order, similar in appearance
to the grasses, but having a solid stem, frequently angular, and
being destitute for the most part of the nutritive qualities for
which grasses are remarkable. The species are generally found
in marshes, ditches, running streams, barren heaths, or on the
seashore. Some of the sedges have edible roots or tubers, and
others are useful for industrial purposes, such as in the manu-
facturing of paper, mats, ropes, baskets, chair-bottoms, &c.
1. Kyllingia, of which New South Wales has K. *intermedia*, K.
monocephala, and K. *cylindrica*, consists of small plants with simple
stems leafy at the base only. K. *monocephala* is very common in
moist shady places, and is easily distinguished by its globular
flower-head.

G

2. Cyperus is a large genus, and includes plants from a few inches to 2 or 3 feet in height. The species are C. *eragrostis*, C. *flavescens*, C. *unilioides*, C. *polystachyus*, C. *pygmæus*, C. *platystylis*, C. *tenellus*, C. *gracilis*, C. *enervis*, C. *debilis*, C. *difformis*, C. *tetraphyllus*, C. *trinervis*, C. *haspan*, C. *concinnus*, C. *filipes*, C. *vaginatus*, C. *dactylotes*, C. *Gilesii*, C. *fulvus*, C. *carinatus*, C. *pilosus*, C. *ornatus*, C. *iria*, C. *rotundus*, C. *congestus*, C. *subulatus*, C. *lucidus*, C. *exaltatus*, C. *hæmatodes*, C. *Bowmanni*, C. *tenuiflorus*, C. *castaneus*, C. *leiocaulon*. Many of these species are common to other parts of the world. C. *difformis* is widely spread over the tropical and subtropical regions of the old world ; and C. *rotundus* (known as nut-grass) is identical with that which has proved so troublesome to sugar plantations in the West Indies, and to gardens in New South Wales.

3. Heleocharis is represented here by H. *sphacelata*, H. *compacta*, H. *tetraquetra*, H. *cylindrostachys*, H. *acuta*, H. *multicaulis*, H. *atricha*, H. *atropurpurea*, and H. *acicularis*. These are marsh or aquatic plants, with simple stems and leafless. The spikelets are solitary and terminal, and the flowers usually have three stamens.

4. Of Fimbristylis, F. *punctata*, F. *nutans*, F. *monostachya*, F. *velata*, F. *æstivalis*, F. *dichotoma*, F. *communis*, F. *ferruginea*, F. *cyperoides*, F. *Neilsoni*, and F. *barbata* occur in different parts of the Colony, principally in the north, only a few species extending to the southern parts of New South Wales and Victoria. They are for the most part small tufted annuals with terminal flowers.

5. Scirpus, including Isolepis, has S. *fluitans*, S. *lenticularis*, S. *crassiusculus*, S. *setaceus*, S. *riparius*, S. *cartilagineus*, S. *inundatus*, S. *prolifer*, S. *nodosus*, S. *supinus*, S. *mucronatus*, S. *pungens*, S. *lacustris*, S. *littoralis*, and S. *maritimus*. These are found in marshes, lagoons, and along the sides of rivers, being common everywhere.

6. Lipocarpha *microcephala* is a tufted annual, only a few inches high. It is plentiful on the Hawkesbury.

7. Fuirena *glomerata* extends from N. Australia and Queensland to the northern parts of New South Wales. It is about a foot high, and has clusters of terminal spikelets.

8. Exocarya *scleroides* is the only species of the genus. It occurs on the Richmond and Clarence rivers, is about 2 feet high, and has flowers in compound umbels.

9. Lepironia *mucronata* rises to the height of 2 or 3 feet, with rush-like stems transversely septate inside.

10. Chorizandra is nearly allied to the last genus. C. *sphærocephala*, C. *enodis*, and C. *cymbaria* have also stems transversely septate inside, but the flowers have the appearance of globular heads of numerous small spikelets.

11. Oreobolus *pumilio* is a dwarf much-branched plant, and forms dense cushion-like leafy tufts on the Australian Alps.

12. Rynchospora *glauca* is common to Queensland and New South Wales. The stem is simple, usually leafy, the spikelets brown and clustered, and the nuts obovate and peaked.

13. Cyathochæte *diandra* has a slender stem of about 2 feet, with a loose panicle of flowers, mostly diandrous.

14. Carpha *alpina* is a perennial plant from a few inches to a foot in height, and limited to alpine localities. It is common to New South Wales, Victoria, and Tasmania.

15. Schœnus is a genus consisting of rigid, nearly leafless plants. The species in various districts are S. *turbinatus*, S. *aphyllus*, S. *imberbis*, S. *ericetorum*, S. *nitens*, S. *Moorei*, S. *villosus*, S. *calostachyus*, S. *brevifolius*, S. *melanostachys*, S. *vaginatus*, S. *apogon*, S. *axillaris*, S. *deustus*, S. *sphærocephalus*, and S. *paludosus*. It is said that all sorts of stock are fond of S. *axillaris*, and that it springs up after the first winter-rains.

16. Lepidospora *tenuissima* is a plant with a creeping, very slender rhizome, from a few inches to a foot. The hypogynous scales are those of Lepidosperma, but in other respects it resembles Schœnus.

17. Lepidosperma may be distinguished by its peculiar hypogynous scales. The stems are usually flat, and the flowers paniculate. Some of the species attain the height of several feet, and they are utilised for their fibre. Many of the following are common to New South Wales and Victoria, viz.:—L. *gladiatum*, L. *exaltatum*, L. *longitudinale*, L. *concavum*, L. *viscidum*, L. *laterale*, L. *lineare*, L. *canescens*, L. *flexuosum*, L. *filiforme*, and L. *Neesii*.

18. Cladium.—This genus consists of tall rush-like plants, varying in habit, some occurring in or near water, and others in dry sandy places. They are C. *mariscus*, C. *insulare*, C. *articulatum*, C. *glomeratum*, C. *tetraquetum*, C. *schœnoides*, C. *Gunnii*, C. *junceum*, C. *teretifolium*, C. *trifidum*, C. *microstachyum*, C. *Sieberi*, C. *melanocarpum*, C. *psittacorum*, C. *asperum*, and C. *xanthocarpum*. In these species the Baron includes Brown's Gahnia and Lampocarya, the former having some tall plants, and both remarkable for their black and red pendulous seeds.

19. Caustis *pentandra*, C. *flexuosa*, C. *recurvata*, and C. *restiacea* are all peculiar to Australia, having sheathing scales similar to those of Restiads and numerous flexuose or straight branchlets. In some places they are made into brooms.

20. Scleria *hebecarpa* and S. *sphacelata* are the only species of the genus in New South Wales. They have triquetrous stems, clustered spikelets, and nuts usually white, and raised on a thickened more or less three-lobed disc.

21. Uncinia differs only from Carex in the hooked bristles projecting from the utricle. U. *riparia* and U. *tenella* are alpine species, common to New South Wales and Victoria. U. *debilior* is peculiar to Lord Howe's Island.

22. Carex, which is the largest genus of the Cyperaceæ, consists of perennials with glass-like leaves. The flowers are unisexual in androgynous spikelets, and the nuts are flattened or three-angled, enclosed in an enlarged utricle. The species are C. *cephalotes*, C. *acicularis*, C. *capillacea*, C. *inversa*, C. *canescens*, C. *echinata*, C. *hypandra*, C. *chlorantha*, C. *paniculata*, C. *declinata*, C. *tereticaulis*, C. *gracilis*, C. *contracta*, C. *Gaudichaudiana*, C. *acuta*, C. *lobolepis*, C. *pumila*, C. *breviculmis*, C. *Gunniana*, C. *maculata*, C. *Brownii*, C. *longifolia*, C. *pseudocyperus*, C. *neesii*.

Though the sedges are not so nutritious as the grasses, and only a few of them can be regarded as fodder-plants, yet they play an important part in the economy of nature, penetrating into regions where other less-hardy plants cannot flourish, binding and protecting the banks of rivers from the fury of floods, and affording many properties highly useful in medicine and arts.

23. Next to the Cyperaceæ, the GRAMINACEÆ, or Grasses, are in point of numbers the most important order in Australia, about 350 species being recorded from various parts of the continent. As forage-plants the grasses perform a most useful part in the economy of nature, whilst many of them afford food and clothing to man, as well as material for many industrial purposes. With the exception of Lolium *temulentum*, which is not indigenous in Australia, no species of grass is known to possess deleterious properties, and many are much improved by cultivation.

1. Eriochloa is a genus common to the new and old world, and the two species E. *punctata* and E. *annulata* have a wide range in tropical Asia, and though common to Queensland and New South Wales are nowhere very abundant.

2. Paspalum *scrobiculatum*, P. *minutiflorum*, and P. *brevifolium* are grasses confined to moist and shady places. P. *distichum* (Brown's P. *littorale*), sometimes called "water-couch," is a species of rapid growth, and though troublesome in cultivated ground is likely to prove valuable as a fodder-plant.

3. Panicum.—This is a large genus comprising species very different in character and habit, some being found in or near water, and not containing much nourishment, and others being highly prized for their fattening properties, and constituting at certain seasons the principal food for cattle on the stations in the interior. The species for New South Wales are P. *cœnicolum*, P. *divaricatissimum*, P. *macractinum*, P. *sanguinale*, P. *tenuissimum*, P. *parviflorum*, P. *Baleyi*, P. *leucophœum*, P. *semialatum*, P. *flavidum*, P. *gracile*, P. *helopus*, P. *distachyon*, P. *reversum*, P. *crus-*

galli, P. *indicum,* P. *foliosum,* P. *adspersum,* P. *uncinulatum,* P. *repens,* P. *pygmæum,* P. *marginatum,* P. *obseptum,* P. *bicolor,* P. *melananthum,* P. *effusum,* P. *Mitchelli,* P. *decompositum,* P. *trachyrachis,* P. *prolutum,* P. *spinescens,* P. *paradoxum,* and P. *atrovirens.*

4. Oplismenus *compositus* is a weak grass growing in moist places or near water, and differing little from the Panic grasses.

5. Setaria *glauca* springs up on the banks of creeks, and also in cultivation, principally under the shade of corn. Though inferior to the cultivated species, it is a good grass.

6. Pennisetum *compressum* is nearly allied to the last plant, and differing chiefly in its involucre.

7. Cenchrus *Australis* is rare to the south of Port Jackson; but it is more plentiful in the Northern districts, rising occasionally to the height of 9 feet.

8. Spinifex *hirsutus* is a coarse grass, creeping in the sand and forming large tufts. It is abundant on the seashore; but S. *paradoxus* is a smaller plant, found on the Murray and Darling, and extending to Central Australia.

9. Perotis *rara* is a slender grass, ranging from Northern Australia and Queensland to the warmer parts of New South Wales.

10. Arundinella *Nepalensis* is a tall reed-like grass, attaining a height of 8 feet in the Northern districts. Under cultivation it has been cut three times during the season.

11. Tragus *racemosus* (Lappago *racemosa,* Willd.) is a spreading plant, the empty glume of the florets having five prominent nerves armed with rigid hooked bristles. It is a Northern species.

12. Neurachne has three species in New South Wales—N. *alopecuroides,* N. *Mitchelliana,* and N. *Munroi.* They occur in the interior, and the second is regarded as a good winter grass.

13. Zoysia *pungens* is a small grass limited to sandy shores and salt marshes.

14. Imperata *arundinacea,* the "Blady Grass" of the colonists, is a useful fodder-plant when young.

15. Erianthus *fulvus* (Pollinia *fulva,* Benth.) is a Northern species, attaining sometimes a height of 4 feet.

16. Arthraxon *ciliare* is a rare and slender grass known as yet only from New England.

17. Lepturus *incurvatus* and L. *cylindricus* are tufted grasses found in salt marshes or on the sea-coasts, and rare in New South Wales.

18. Hemarthria *compressa* is a grass with decumbent or creeping stems, and occurring in moist places.

19. Ischæmum *triticeum,* I. *Australe,* and I. *pectinatum* are fine growing grasses in Queensland, but I. *ciliare,* which is peculiar to New South Wales, occurs only sparingly on the Hunter.

20. Andropogon, including Sorghum, is well represented, and comprises some of the most nutritious grasses in the Colony. They are A. *erianthoides*, A. *sericeus*, A. *affinis*, A. *pertusus*, A. *intermedius*, A. *bombycinus*, A. *refractus*, A. *lachnantherus*, A. *contortus*, A. *Gryllus*, A. *montanus*, A. *Halepensis*, and A. *Australis*. Of these, A. *Halepensis* is the most widely distributed throughout the world; and, according to Baron Mueller, "it yields a large hay crop, as it may be cut half-a-dozen times in a season." "But," he adds, "it will mat the soil with its deep and spreading roots; hence it should be kept from cultivated fields."

21. Anthistiria includes what are called "the Kangaroo grasses." A. *ciliata*, the commonest in New South Wales, is an excellent grass, and the chemical analysis of it gives—albumen, 2·05; gluten, 4·67; starch, 0·69; gum, 1·67; sugar, 3·06 per cent. (F. v. M. and L. Rummel). A. *avenacea* is a coarser grass, and produces a large amount of fodder. A. *membranacea* is peculiarly suited for dry hot regions.

22. Apluda *mutica* was collected by Leichhardt in some part of New South Wales, but Mr. Bentham scarcely considered it indigenous.

23. Alopecurus *geniculatus* is a glabrous grass with a densely crowded spike of florets. It is common in the western interior.

24. Leersia *hexandra* has not been found near Port Jackson since the days of Brown. It is a weak glabrous plant, attaining several feet, and often rooting in the mud at the lower nodes.

25. Potamophila *parviflora* is an aquatic grass, from 3 to 5 feet. The genus is limited to a single species endemic in Australia.

26. Ehrharta.—E. *juncea* (Tetrarrhena *juncea*, R. Brown) is a long and slender species, the florets having four stamens. It occurs at Mossman's Bay, &c. E. *stipoides* (Microlœna *stipoides*, R. Brown) is a tender and nutritious grass, but it cannot bear much exposure to the sun.

27. Hierochloe *redolens* and H. *rariflora* are alpine species, common to Tasmania, Victoria, and the southern parts of the Colony.

28. Aristida has the following species in Queensland and New South Wales:—A. *stipoides*, A. *arenaria*, A. *Behriana*, A. *leptopoda*, A. *vagans*, A. *ramosa*, A. *calycina*, and A. *depressa*. These are for the most part harsh grasses, capable of enduring a great amount of heat, but not so nutritious as many others, whilst the terminal trifid awn of the flowering glume is injurious to sheep.

29. Stipa is a genus of a varying character. Some of the species are ornamental, some afford tender herbage, and others are exceedingly troublesome to sheep-farmers on account of their awned seeds, which are not only injurious to the wool, but fatal to the sheep. The species are S. *elegantissima*, S. *Tuckeri*, S. *verticillata*, S. *flavescens*, S. *setacea*, S. *semibarbata*, S. *pubescens*, S. *aristiglumis*, and S. *scabra*.

30. Dichelachne *crinita* and D. *sciurea* are very frequent in different districts. They were referred by R. Brown to the genus Agrostis.

31. Pentapogon *Billardieri* derives its generic name from the five lobes or awns at the end of the flowering glume. P. *Billardieri* is a southern species.

32. Echinopogon *ovatus* is found generally by the sides of creeks or in moist places. The heads of florets are ovoid-globular, and have a bristly appearance.

33. Amphipogon *strictus* is erect and slender, growing in sandy and damp places, but not of much utility for forage.

34. Pappophorum *commune* is remarkable for its broad membranous glume, with numerous nerves, ending in plumose awns. There are several varieties of this species differing in size and colour.

35. Sporobolus has the following species:—S. *indicus* (a hard tufted grass widely distributed); S. *virginicus*, common near the sea-coast; and S. *diander*, S. *pulchellus*, S. *Lindleyi*, and S. *actinocladus*, principally beyond the Dividing Range.

36. Agrostis (including Deyeuxia, Clarion) is represented by A. *Muelleri*, A. *scabra*, A. *venusta*, A. *Solandri*, A. *montana*, A. *quadriseta*, A. *frigida*, A. *nivalis*, and A. *breviglumis*. Some of these are valued for pastures, but they are not plentiful.

37. Aira *cæspitosa* is a tall perennial, growing in moist places in the Southern districts.

38. Trisetum *subspicatum* is an alpine species, varying from 6 inches to above 2 feet in height.

39. Eriachne *aristidea* and E. *obtusa* occur in the interior. The flowering glumes are silky-hairy.

40. Anisopogon *avenaceus* is a tall oat-like grass, extending from Port Jackson to the Blue Mountains.

41. Danthonia.—The species (D. *paradoxa*, D. *bipartita*, D. *carphoides*, D. *penicillata*, D. *robusta*, D. *pauciflora*, and D. *nervosa*) are difficult to define, as they seem to run into each other. They are nutritive, and valued as pasture. D. *nervosa* is Amphibromus *Neesii* of Steudel.

42. Astrebla *pectinata* and A. *triticoides*, known as "Mitchell grasses," occur from the Darling to the Barrier Range and Cooper's Creek. Mr. Bailey, F.L.S., describes them as perennial desert grasses, resisting drought, and possessing fattening qualities.

43. Cynodon *dactylon*, the well-known "Couch-grass" of the colonists, is one of our most valuable pasture grasses, especially on this side of the Dividing Range. The chemical analysis gives—albumen, 1 60; gluten, 6·45; starch, 4·00; gum, 3·10; and sugar, 3·60 per cent.

44. Chloris *acicularis*, C. *truncata*, and C. *ventricosa* resemble the larger forms of Cynodon, but the spikes of florets are more numerous and spread horizontally.

45. Eleusine *Ægyptiaca*, E. *Indica*, E. *digitata*, and E. *Chinensis* occur principally in the interior, and are characterised by their flat spikelets, imbricate in two rows along one side of the digitate branches of a simple panicle.

46. Poa is a genus which includes some of our most valuable grasses. The species are P. *cæspitosa*, P. *nodosa*, P. *lepida*, P. *Fordeana*, P. *fluitans* (Glyceria *fluitans*, R. Br.), P. *latispicea*, and P. *ramigera*. P. *cæspitosa* is very hardy, and, though not generally liked by cattle, forms some of their principal food in times of drought.

47. Festuca *duriuscula* is one of the widely-dispersed forms of the "Sheep's Fescue"; F. *littoralis* grows only on the sandy shore; and F. *Hookeriana* is a stout perennial alpine species.

48. Diplachne *loliiformis* is found in the northern and western parts of the Colony, and D. *fusca* from the Lachlan to the Darling.

49. Triodia *Mitchelii*, T. *irritans*, and T. *microdon* are the troublesome prickly grasses of the desert, sometimes called "Porcupine grasses," but more generally "Spinifex," though in no way connected with that genus.

50. Distichlis *maritima* is a rigid, glabrous, much-branched grass, forming low leafy tufts on the southern parts of the sea-coast.

51. Bromus *arenarius* is valued for its fattening qualities by stockholders beyond the Dividing Range.

52. Eragrostis is a genus nearly allied to Poa, and has the following species, many of which are highly useful for pasture:— E. *tenella*, E. *nigra*, E. *megalosperma*, E. *pilosa*, E. *leptostachya*, E. *diandra*, E. *Brownii*, E. *laniflora*, E. *eriopoda*, E. *setifolia*, E. *lacunaria*, and E. *falcata*.

53. Ectrosia *leporina* resembles Triraphis *mollis*, having a panicle of dense flowers with fine awns. The spikelets have one or two fertile flowers with empty glumes above them. The species is limited to the north-western part of the Colony.

54. Elythrophorus *articulatus* is indigenous on the Murray and in the interior.

55. Trirhaphis *mollis* has flowering glumes with three narrow lopes tapering into awns. It occurs on the Narran and Darling.

56. Agropyron includes R. Brown's Triticum *scabrum*, now A. *scabrum*, common from Port Jackson to the Blue Mountains, and A. *velutinum* and A. *pectinatum* to Tasmania, Victoria, and the southern parts of the Colony.

57. Arundo *phragmites* is a species of reed found in wet ditches, marshes, and shallow water, almost everywhere. It sometimes

attains a height of 6 feet, and has long creeping roots. Since the early days of the Colony many of the native grasses, partly owing to the cultivation of cereals and partly to the naturalisation of foreign grasses, have become less frequent in the Settled Districts. Amongst introduced grasses may be mentioned Stenotaphrum *Americanum* (Schrank.), Phalaris *Canariensis* (Linn.), P. *minor* (Retz), Anthoxanthum *odoratum* (Linn.), Holcus *lanatus* (Linn.), Avena *fatua* (Linn.), Dactylis *glomerata* (Linn.), Poa *annua* (Linn.), P. *glauca* (E. B.), P. *pratensis* (Willd.), Briza *major* (Linn.), B. *minor* (Linn.), Bromus *sterilis* (Linn.), B. *mollis* (Linn.), Ceratochloa *unioloides* (D. C.), Lolium *temulentum* (Linn.), L. *perenne* (Linn.), Hordeum *nodosum* (Linn.), H. *murinum* (Linn.), and Aira *præcox* (Linn.) Two grasses, viz., Cynodon *dactylon* and Paspalum *distichum*, which in the days of R. Brown were probably found only near the coast, have spread widely through the Colony, the former being the most plentiful of the grasses from Port Jackson to the Blue Mountains, and the latter having established itself on alluvial soil in some places to the impediment of agriculture.

Of Monocotyledonous plants for New South Wales the numbers are :—

	Gen.	Sp.
I. Calyceæ Perigynæ	52	164
II. do. Hypogynæ	67	158
III. Acalyceæ do.	79	343
	198	665

(III.) ACOTYLEDONEÆ.

The plants of this division of the vegetable kingdom have no stamens, pistils, or seeds, and the reproduction is carried on by what are termed spores. The higher orders of these cryptogams, that is such as have true stems enclosing bundles of vascular tissue and the spores enclosed in spore cases, are,—

    1. RHIZOSPERMÆ.
    2. LYCOPODINEÆ.
    3. FILICES.

I. The RHIZOSPERMS, or Pepperworts of Lindley, are stemless plants, creeping or floating, and in New South Wales they are limited to a few species. Azolla *pinnata* and A. *filiculoides* (A. *rubra*, R. Br.) are small floating plants with branching and rooting leafy stems, covering sometimes the surface of ponds and lagoons. A. *pinnata* occurs principally to the north, and A. *filiculoides*, which is rather a smaller plant, is plentiful from Paterson's River to Victoria.

Marsilea *quadrifolia* (the "Nardoo" of the blacks) is common to all the Australian Colonies, and as it varies considerably in proportion to its proximity to water some botanists have divided the different forms into distinct species. When growing in water the stems are long and the leaflets large and smooth; but when the water recedes the plants become smaller in every respect. In the interior the fronds assume a rusty-villous appearance.

Pilularia *globulifera* has a filiform rhizome, creeping and rooting at the nodes. The barren fronds are of a bright-green colour, 2 or 3 inches long. This species is found in most of the Australian Colonies and Tasmania.

II. LYCOPODINEÆ, or "club mosses," are usually moss-like plants, with creeping stems and imbricated leaves, and, according to Lindley, they are "intermediate, as it were, between ferns and conifers on the one hand, and ferns and mosses on the other."

1. Psilotum *triquetrum* grows from the crevices of rocks or forks of trees, and has erect or pendulous stems, repeatedly dichotomous, and minute scale-like leaves. 2. Tmesipteris *Tannensis* has simple leafy stems, varying from 6 inches to a foot. It grows frequently on the caudices of tree-ferns. 3. The genus Lycopodium, which has small leaves, usually in four rows inserted round the stem, has the following species in New South Wales:— L. *selago*, L. *clavatum*, L. *Carolinianum*, L. *laterale*, and L. *densum*, the last two occurring in the neighbourhood of Sydney. L. *densum*, which has the appearance of a dwarf conifer, is an elegant plant rising to the height of 2 feet or more. 4. Selaginella, which differs from Lycopodium chiefly in habit, has S. *Pressiana*, S. *uliginosa*, and S. *Belangeri*. 5. Phylloglossum *Drummondii* is a small stemless plant with a few fibrous roots. It is found in damp places in the southern parts of the Colony.

III. The FILICES or ferns are plants of a higher kind than the preceding order, having a rhizome, or stem, or trunk, sending forth leaflike fronds. The fructification consists of spore-cases, usually small and collected in clusters, called sori, which are either naked or covered with a membrane termed an indusium. The manner in which the sori are placed on the fronds, and the presence or absence if an indusium, afford characteristics whereby the species can be classified. Ferns, which are abundantly dispersed throughout the world, diminishing, however, in numbers in dry or cold countries, are amongst the most elegant and graceful of plants, delighting the eye by their external appearance, and affording by their minute structure endless material for anatomical investigation. The number of British ferns is limited to about forty species, and these, with the exception of Osmunda *regalis*, which attains the height of several feet, are comparatively small plants, the tree and climbing ferns being most frequent in the humid forests of tropical islands, and yet extending to

Tasmania and New Zealand in latitude nearly 40 degrees south. Though in the interior of Australia ferns cease to exist, and in many large districts they are confined to a few species, the Australian ferns, nevertheless, scattered as they are abundantly on the mountain ranges, in the deep gullies, or in the damp shady scrubs of the eastern coast, are reckoned by Baron Mueller as exceeding 200 species. Of these, only eleven are found in Western Australia, whilst it may be remarked that no tree-fern has been discovered in that colony. Following the classification of Mr. Bentham and Baron Mueller, the ferns of New South Wales may be distributed in six tribes.

1. Ophioglosseœ.—In this tribe the fronds are not circinate, the barren one leaflike, the fertile spike-like, and the spore-cases globular, two-valved, without any ring, and sessile in two rows, or in small clusters on the spike or its branches. Ophioglossum *vulgatum*, or "Adder's Tongue," is a small fern only a few inches high; but O. *pendulum*, which hangs from trees and rocks, is ribbon-like, and sometimes many feet in length. Botrychium *lunaria* and B. *ternatum* have pinnate fronds, the fertile ones forming a kind of panicle. They seldom exceed many inches in height, the latter being larger and more divided.

2. Marattieæ have the fronds circinate in vernation, the spore-cases without any perfect ring, opening in two valves or a longitudinal slit. Lygodium *scandens* is a climbing fern with long twining stems. The fronds are pinnated, and the spore-cases are sessile bordering the pinnules.

Schizæa *rupestris* seldom exceeds 4 inches, is erect in habit, and has the sori closely imbricate at the end of the fertile branches. S. *dichotoma* rises to the height of a foot or more, and the fronds are dichotomously divided into numerous branches. Marattia *fraxinea* is a handsome species with fronds of several feet. It is common to Norfolk and Lord Howe's Islands, as well as Queensland.

3. Osmundeæ include species with fronds circinate in vernation, divided or compound, and spore-cases globular, without any perfect ring. Ceratopteris *thalictroides* is an aquatic fern, with fronds 3-pinnately divided, the revolute margins of the pinnæ enclosing the fructification. Gleichenia (including Brown's Platyzoma) has the species G. *platyzoma*, G. *circinata*, G. *dicarpa*, G. *flabellata*, and G. *Hermanni*. The fronds are erect or scrambling, the main rachis dichotomous (except in G. *platyzoma*), the sori without indusium, and surrounded by an obscure transverse ring opening vertically in two valves. Osmunda (Todea, Willd.) includes O. *barbara*, a large tree-like fern growing in or near water; O. *Fraseri*, one of the most delicate and graceful of Australian ferns, found in deep gullies on the Blue Mountains; and O. *Moorei*, from Lord Howe's Island.

4. Hymenophylleæ.—In this tribe the fronds are thin and membranous, usually small, and the spore-cases depressed, with a transverse ring sessile or nearly so, on a columnar receptacle arising from the base of a cup-shaped or deeply two-valved indusium.   Of Trichomanes, the following are peculiar to New South Wales or Queensland :—T. *vitiense*, T. *parvulum*, T. *digitatum*, T. *venosum*, T. *rigidum*, T. *humile*, T. *caudatum*, and T. *Bauerianum*. Hymenophyllum differs from Trichomanes in having the indusium in the exserted part divided into two broad lodes or valves.   The species are H. *marginatum*, H. *nitens*, H. *Javanicum*, H. *multifidum*, H. *bivalve*, and the cosmopolitan H. *Tunbridgense*.

5. Cyatheæ comprise species with aborescent trunks, large fronds twice or thrice pinnate, and numerous spore-cases, with a more or less oblique ring in globular sori on the under surface of the segments.   Cyathea (including Hemitelia) has C. *Lindsayana*, C. *Macarthuri*, C. *medullaris*, C. *brevipinnea*, and C. *Moorei*, varying in height from 10 to 50 feet, with fronds in one species from 10 to 20 feet in length.   None of these occur near Sydney, but in the genus Alsophila, which differs from the preceding in having sori without any indusium, A. *Australis* and A. *Leichhardtiana* are found near the coast, as well as on the Blue Mountains ; A. *Cooperi* at Illawarra and the Kurrajong ; and A. *Loddigesii* at Cape Byron.   A. *excelsa*, the loftiest of the genus, has its typical form in Norfolk Island, and rises to the height of 60 feet.

6. Polypodieæ.—In this large tribe, which includes a vast variety of forms, from the minute Grammatis *leptophylla* to the tree-fern Dicksonia *Billardieri*, the spore-cases are small, and have a longitudinal ring,   These are generally on the under surface of the frond, with or without an indusium.

Dicksonia.—This genus includes the fine tree-ferns D. *Billardieri* (D. *antarctica*, Labill.) and D. *Youngiœ*, as well as the tender D. *davallioides* and D. *nephrodioides* (Deparia, Baker) from Lord Howe's Island.   The first of these rises to 30 feet and upwards, and being a hardy species it can easily be cultivated.

Davallia *pyxidata* and D. *dubia* are widely distributed, the first being remarkable for its scaly rhizome, and the other for its large fronds, resembling Dicksonia *davallioides*, but much more rigid in texture.

Vittaria *elongata* has linear fronds sometimes 2 feet in length. The sori are continuous along the intramarginal vein with a two-valved indusium.

Lindsaya has the species L. *linearis* and L. *microphylla* from Port Jackson to the Blue Mountains, L. *trichomanoides* from the latter locality, and L. *incisa* from Clarence River.   The texture of the fronds is tender, and the pinnules small.

Adiantum, known popularly as "Maidenhair," has A. *æthiopicum*, A. *formosum*, A. *affine*, A. *diaphanum*, and A. *hispidulum* in many parts of the Colony. Some forms of A. *æthiopicum* approach very near the true "maidenhair" of Europe.

Cheilanthes (including Nothoclœna) is now restricted to C. *vellea*, C. *distans*, and C. *tenuifolia*. The fronds are small, and the last, in different forms, is the commonest of our ferns, extending even to Central Australia.

Pteris has in New South Wales, not only a variety of the widely-ranging P. *aquilina* (the common "bracken" of Europe), but also P. *geraniifolia*, P. *paradoxa*, P. *falcata*, P. *longifolia*, P. *umbrosa*, P. *arguta*, P. *incisa* ("bat's-wing"), P. *comans*, and P. *marginata*.

Lomaria is a genus distinguished for the great difference between the fertile and barren fronds, the former becoming contracted, and the sori covering almost the whole of the under surface. L. *Patersoni*, L. *discolor*, and L. *Capensis* are common near Sydney; L. *fluviatilis* and L. *alpina* are Southern species; whilst L. *attenuata* and L. *Fullageri* are indigenous in Lord Howe's Island. Blechnum *cartilagineum*, B. *lævigatum*, and B. *serrulatum* are very like species of Lomaria, but in this genus the sori are in a continuous line on each side of the midrib without covering the whole under surface.

Woodwardia (Doodia, R. Br.) is a variable genus, and the fronds are usually scabrous. The species, according to the Baron, are only two—W *aspera* and W. *caudata*—and in different localities they produce many varieties. Asplenium as a genus is characterised by linear sori on veins proceeding from the midrib, and having an indusium attached along one side of the vein, and opening along the other. Many of the species are much admired, and obtain a place in conservatories and greenhouses. The smaller ones are A. *trichomanes*, A. *flabellifolium*, A. *attenuatum*, A. *marinum*; the more showy, A. *nidus*, A. *falcatum*, A. *Hookerianum*, A. *furcatum*, A. *bulbiferum*, A. *pteridioides* (from Lord Howe's Island), A. *umbrosum*, A. *maximum*, A. *melanochlamys*, A. *Robinsonii*, and A. *diversifolium*. The genus, as now arranged, includes Allantodia and Diplazium.

Aspidium, with which Nephrodium and Nephrolepis are united, has the following:—A. *cordifolium*, A. *ramosum* (a fern creeping up trees to a great height), A. *unitum*, A. *molle*, A. *truncatum*, A. *aculeatum*, A. *aristatum*, A. *capense*. A. *decompositum*, A. *tenerum*, A. *uliginosum*, and A. *hispidum*. A. *decompositum* has several varieties, and one, from the fact of its being seldom found with any indusium, is sometimes referred to Polypodium.

Polypodium is a large genus, and the species differ very much in habit, some being erect, some creeping, and others climbing. They are generally recognised by the absence of any indusium.

Those found in different parts of the Colony are P. *Australe*, P. *Hookeri* (from Lord Howe's Island and Queensland), P. *grammitidis*, P. *tenellum*, P. *proliferum*, P. *serpens*, P. *confluens*, P. *attenuatum*, P. *pustulatum*, P. *scandens*, P. *rigidulum*, and P. *punctatum*.

Hypolepis *tenuifolia* is a fern resembling Polypodium *punctatum*, but the fronds are usually delicate in texture, and the sori differently inserted.

Grammitis *rutifolia* is a small hairy fern common to the southwest of Chili, New Zealand, and Australia.

G. *leptophylla*, another minute and delicate species, is widely dispersed over the temperate and subtropical regions of the old world, and also in the Andes. It has been collected at Port Stephens, Cowra, and beyond Mudgee, but seems rare, and with the preceding occurs in Western Australia.

Acrostichum has its sori confluent, covering the under surface of the fertile fronds. P. *aureum* and A. *spicatum* are frequent in Queensland, and extend to the Northern districts of New South Wales.

Platycerium has large fronds, the veins prominent and reticulate, and the sori forming large patches towards the ends of the fronds. P. *alcicorne* is common on trees and rocks from Port Jackson to the Blue Mountains. P. *grande*, a large species, does not extend so far south.

From the review of Acotyledoneæ it appears that the genera and species are—

				Gen.	Spec.
1.	Rhizospermæ	...	...	3	4
2.	Lycopodineæ	...	...	5	11
3.	Filices	...	...	28	125
				36	140

The whole number of vascular plants now known to exist in New South Wales is 3,154, being more than a third of those systematically known for all Australia.*

---

* The following species, which are not recorded in the previous pages, have been added only recently to the Baron's Census :—

Loranthus *grandibracteus*
Pomaderris *phylicifolia*
Helmholtzia *glaberrima*
Montia *fontana*
Polygala *Chinensis*
Agonis *Scortechini*
Ipomæa *plebeia*
I. *heterophylla*
I. *aquatica*
I. *cataractæ*
Hedraianthera *porphyropetala*
Ficus *stenocarpa*

Boronia *pilosa*
B. *rhombifolia*
Sida *Spenceriana*
S. *platycalyx*
Tribulus *minutus*
Flaveria *Australasica*
Cassia *pumila*
Aponogenton *monostachyus*
Glycine *Latrobeana*
Schizæa *fistulosa*
Catanospora *Alphandi*
Oncinocalyx *Betchii*.

LIST OF NATURAL ORDERS.

INDEX OF GENERA.

H

116 INDEX.

Sydney : Thomas Richards, Government Printer.—1885.

Printed in the United States
By Bookmasters